CHARLIE –

A JOURNEY FROM THE AUSSIE BUSH TO BATTLE SKIES FAR AWAY

Charles Hume Baldwin
at enlistment in Royal Australian Air Force
Brisbane, 7 December 1941

Cheryl Baldwin 2022

Charlie
A journey from the Aussie bush to battle skies far away

First published in Australia by Cheryl Baldwin ©2022
Copyright © Cheryl Baldwin 2022 Text and images
All Rights Reserved

A catalogue record for this
book is available from the
National Library of Australia

ISBN: 978-0-6452586-1-5 (pbk)

Cover design by Julia Pearson and Cheryl Baldwin © 2022

Typesetting and design by Publicious Book Publishing
Published in collaboration with Publicious Book Publishing
www.publicious.com.au

No part of this book may be reproduced in any form, by photocopying or by any electronic or mechanical means, including information storage or retrieval systems, without permission in writing from both the copyright owner and the publisher of this book.

In memory of Charlie –
Charles Hume Baldwin
1919 – 2005

RAF 550 Squadron Memorial Window,
St Denys Church, North Killingholme, Lincolnshire UK

CHARLES HUME BALDWIN RAAF 420806
KEY DATES OF SERVICE DURING WWII

1941 01 July	Royal Australian Air Force Reserves, Newcastle, NSW
07 Dec	RAAF Sandgate, Q
1942 23 June	RAAF Narromine, NSW
1942 06 Oct 1942 – 1943	RCAF: Brandon, Manitoba Paulson, Manitoba Trenton, Ontario Malton, Ontario
1943 15 July 1943 - 1945	RAF UK West Freugh, Scotland Lichfield, Staffordshire Hixon, Staffordshire Lindholme, Yorkshire Elsham Wolds, Lincolnshire North Killingholme, Lincolnshire Finningley, Yorkshire
1945 28 July	Disembarkation Brisbane, Australia
1946 18 Mar	Demobilisation

RAF Kemble

CONTENTS

Key Dates	iii
Introduction: Charles Hume Baldwin 1919 – 2005	i
Chapter 1: Life on the Land: 1791 - 1938	1
Chapter 2: 1941 RAAF Reserve	6
Chapter 3: RAAF Sandgate	10
Chapter 4: Narromine, NSW 1942	14
Chapter 5: Voyage Across the Pacific 1942	20
Chapter 6: Canada November 1942	23
Chapter 7: Land of Snowy Winters 1942- 43	27
Chapter 8: Arrival UK July 1943	32
Chapter 9: An Accident 9 February 1944	37
Chapter 10: Convalescence February - June 1944	44
Chapter 11: Farewell Fred	47
Chapter 12: Return to Flying 1944	52
Chapter 13: The Lancaster	56
Chapter 14: Finishing School	62
Chapter 15: 550 Squadron October 1944	68
Chapter 16: Battle Skies	74
Chapter 17: Christmas 1944	77
Chapter 18: The Odds	82
Chapter 19: Lost	89

Chapter 20: Post War Life 1946	93
Chapter 21: Peace	100
Chapter 22: Legacy	104
Postscript: A Letter from Cyril Fradley, Hixon UK, 1951	109
Appendix	119
Acknowledgements	122
Resources	129

INTRODUCTION
CHARLES HUME BALDWIN 1919 - 2005

Inspiration to write about my father, Charles Hume Baldwin, known as Charlie came from a photo arriving in the post from my friend Hazel, who lives in Gloucestershire, UK. The photo depicted an embroidered panel, one of ten which were sewn together to form a tapestry. This was the project undertaken by the Gloucestershire County to commemorate the Centenary of the Royal Air Force (RAF) 1918 - 2018.

Each panel had been assigned to a group of women whose nimble fingers stitched and sewed the panels together, completing the tapestry in little more than a year, in time for the celebrations. Hazel had been one of the dedicated band of women working on the panel for RAF Kemble, located five miles south-west of Cirencester. This panel featured the RAF Aerobatic Team, Red Arrows.

Beginning in April 2018, commemorative events and displays were scheduled throughout the year in numerous locations across the UK. These events would have been of great significance to Charlie, who had passed away in March 2005.

After Charlie joined the RAAF in 1941, his training began in Australia and continued in Canada before arriving in the UK, serving with the RAF during WWII.

He served in Bomber Command's 550 Squadron at North Killingholme, Lincs. (Lincolnshire), where an annual reunion is held by a diminishing number of veterans, their families and friends. Local villagers are involved in commemoration events and maintain a small museum in the village hall. It features an archive of the squadron's history, memorabilia and photos.

Had Charlie been alive and well enough to make the long journey to attend a 550 Squadron reunion, he'd also be sure to include a visit to the International Bomber Command Centre, to be officially opened on 12 April 2018.

Another special event would be the service to commemorate the Battle of Britain, at Westminster Abbey on Sunday 16 September 2018, a poignant occasion, as the date would have been Charlie's ninety-ninth birthday. In the Abbey's RAF Chapel, a magnificent stained-glass window features the insignia of squadrons which took part in the Battle of Britain.

With a desire to tell the story of Charlie's journey into war and his return home to Australia, I decided to write it as a tribute to his life and a record for his family.

Charlie was born on 16 September 1919 at Glen Innes, a town on the Northern Tablelands in the New England region of New South Wales, Australia. He was the third of seven children, five daughters and two sons born between 1916 and 1929 to Glen Adair Baldwin and Ann, known as Annie (nee Salmond).

CHAPTER 1

Life on the Land: 1791 - 1938

For Charlie, a love of the land began with childhood, and a life of living and working on the land can be traced back to his English-born, great-great-grandfather, Henry Baldwin (1769-1843). Each generation between Henry and Charlie would be linked to a rural life.

Henry Baldwin, born in Hertfordshire England, a Yeoman (farmer) was convicted for burning haystacks as a protest against Land Enclosures. Aged 22, he arrived in Sydney, Australia on the Third Fleet ship Admiral Barrington in 1791. Freed in 1797, Henry bought land on the banks of the Hawkesbury River at Freeman's Reach, located 65 kilometres north west of Sydney. He named his property Wellow Farm, where Henry soon prospered as a wheat producer. His farming experience and industrious efforts were rewarded in 1803 with a land grant of 100 acres adjoining his farm.

Two of his sons, Otto and Harvest (Charlie's great-grandfather) followed in the path of early explorers when they were the first to take cattle across the Liverpool Range and squatted on land beside the Namoi River, which flowed across the Liverpool Plains.

Their nephew, Charles (Edwin Charles Baldwin, who was a great-uncle to Charlie, Charles Hume Baldwin) worked with them to establish a shorthorn cattle stud with stock imported from England. After Charles inherited the property, he became well-known for racehorse breeding with thoroughbreds sired by imported Arab stallions.

A rural life for many of Henry's descendants would continue, working as farmers and drovers. Charlie's father, Glen was working on the sheep station, Shannonvale near Glen Innes, during which time, five of Glen and Annie's seven children were born: Nellie, Bonnie, Charlie, Elwyn, and Phyllis. From there the family moved to the rural area of Woodenbong, close to the Queensland border and the two youngest children, Ena and Jacqueline (known as Johnnie) were born. In 1931 the family moved across the border into SE Qld, where they bought a block of land at Innisplain, located near Tamrookum within the Beaudesert District.

Glen and Annie worked hard to build a house and establish the farm they named Merton. The rough wooden house had a kitchen, dining room and four bedrooms and although not large, would accomodate the family quite well.

This was a relief after the cramped conditions they'd previously been living in. The laundry was a separate building at the back of the house and a large tin tub was placed in front of the kitchen fire for bathing. Life wasn't easy, but Annie used to say that if you had a good bed to sleep in and plenty of food on the table, you were well off. When seasons were good, the land was green and productive, but without rain, sparsely covered earth and spindly trees indicated the lean times. In time, part of the property would be irrigated with water from the Logan River to grow lucerne.

The children, all helped with chores around the farm. Nellie, the eldest worked alongside her father, while Bonnie, the second child helped her mother with the household and caring for the younger siblings. There were paddocks to be ploughed, crops to be planted, cows to be milked, pigs and chickens to be fed. On frosty mornings, Charlie would warm his feet on the grass, where the cows had been lying down until he got them up and moving to the bails. From the time he was knee-high to a grasshopper, he could ride a horse, as did his sisters and brother. Without a saddle, the horse carried the four eldest children to school.

In 1938 Glen died of a heart attack, leaving Annie with the three youngest children at home. Eleven-year old Ena and nine-year old Johnnie, were still attending school, with Phyllis having completed school at 14, the school-leaving age. Nellie, now aged 22, the eldest in the family, had been in domestic service on a local Tamrookum property since she turned 14.

At the time, Bonnie was a nanny and housemaid on a property in the western district of Mungallala. She must have been homesick and finding the situation quite hard, judging from her father's words of encouragement in a letter he wrote to her the year before he died:

'Well, Dear old thing, keep a stiff upper lip & hang on, you will get accustomed to it.'

Nellie, Bonnie, Charlie & Elwyn on horseback at *Shannonvale*, Glen Innes, NSW 1923/1924

Glen Adair Baldwin is buried at All Saints Anglican Church, Tamrookum

There wasn't much news from home and paper was scarce, he wrote, but there had been rain the night before, which was a miracle. At that time he was seriously ill, and unable to work the farm, which had to be sub-let, with the family of five moving to a cottage near Tamrookum.

Charlie, after leaving school at 13 had helped on the farm until 1936, when aged 17, he and Elwyn, aged 14 went to work on a sheep station, Trafalgar near Dirranbandi in southwest Qld. Charlie, Elwyn and Bonnie returned home for their father's funeral, helping Annie and the younger children to move back to their farm Merton.

After his father's death, Charlie returned to Trafalgar at Dirranbandi, but then left in 1940 to return home and help operate the family farm at Innisplain.

Charlie (left) and a cousin (identity uncertain) in Sydney before Charlie enlisted in the RAAF Reserves in Newcastle 1941

CHAPTER 2

1941 RAAF Reserve

In 1939 war had been declared on the other side of the world and by the end of 1941, Annie would farewell her elder son, this time to begin training in the RAAF. Charlie, aged 22 years, was a tall, quietly spoken young man, who was going to leave his life on the land and take to the skies. He explained:

> *I was keen to fly and joining up was the thing to do.*

In June 1941, while he was on a holiday visiting his uncle in Newcastle, Charlie enrolled in the RAAF Reserves at No. 2b Mobile Unit. Without his father to talk it over with, Charlie would have discussed his decision with his uncle and the cousin closest to him in age, who may also have been planning to join up.

On the enrolment form, dated 1 July 1941, Charlie was recorded as being a British Subject, born on 16 September 1919, of single status, his religion as C. of E. (Church of England). His place of residence was Innisplain, Beaudesert, Queensland, where he was a Station Hand, also referred to as Jackaroo on his official

Airman's Service record. Importantly, Charlie had no civil convictions or dismissals from His Majesty's Service.

He took the oath: *"to well and truly serve Our Sovereign Lord the King as a member of the Air Force Reserve of the Commonwealth of Australia, and that I will resist His Majesty's enemies and cause His Majesty's Peace to be kept and maintained, and that I will in all matters appertaining to my service faithfully discharge my duty according to the law. So help me God!"*

The Medical Certificate dated 19/9/41, hand-written by the Examining Medical Officer, whose scrawl of a signature was illegible, had certified Charlie as medically suitable for the Air Force Reserve. Physically, he was 6ft tall, with medium complexion, brown hair and blue eyes, both with 6/6 vision and no colour vision problem. He weighed 168lb. with a chest measurement of 36/39 inches.

His decision to enlist would change his life in ways he couldn't have foreseen. From his life of hard work and long days in the bush, droving horses or sheep, often alone with his horse and the livestock in his care, he'd set out on a new journey that would take him to faraway places with different landscapes, climates and challenges.

A generation earlier, two of Charlie's uncles, Charles Dudley Baldwin and Jack Gordon Baldwin left their rural life to go to war. Both experienced horsemen, they joined the Australian Imperial Force (AIF) in WWI. They served together in the 2nd Light Horse Brigade, 7th Regiment and returned from war in 1919, the year Charlie was born. In choosing the RAAF and his dream to fly, Charlie didn't follow in his uncles' footsteps to join the AIF.

Once he'd signed up with the RAAF Reserves, Charlie was impatient, waiting to be called up for training, but he would have been busy studying the manual given to Air Crew Reservists. Based on RAF training manuals, it provided an introduction on all aspects of flying, including aerodynamics, navigation, meteorology and aircraft engines. For Charlie and others with schooling that finished at the age of 14, it would have been a challenge to equip themselves with this new complex knowledge, but necessary if they were to past the tests during training, both in the classroom and in the air.

During this period of waiting Charlie took his religious confirmation on 21 September at the All Saints Church of England, Tamrookum, where his father had been buried in the church graveyard. The occasion was marked by the presentation of a tiny red book, *Helps to Worship: a manual for Holy Communion and Daily Prayer*.

At the back of Charlie's little red confirmation book, which he left in the care of his mother, Annie would later record important dates of his RAAF service. On the end page, her distinctive handwriting with large inky loops, listed the dates of Charlie's arrival and departure at each place he'd been posted to, beginning with his RAAF training in Australia to Canada, where he trained under the Royal Canadian Air Force (RCAF) and finally his arrival in England on 24 July 1943. This was the last entry with half the page left blank and I was curious to know if Annie had continued with her records in another book.

*CHARLIE – a journey from the Aussie
bush to battle skies far away*

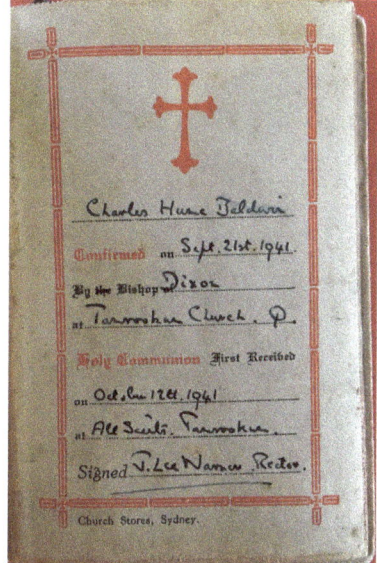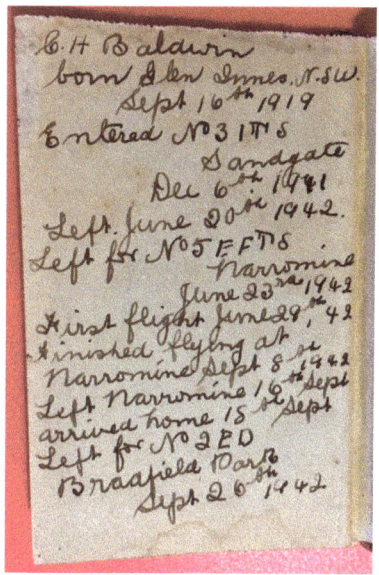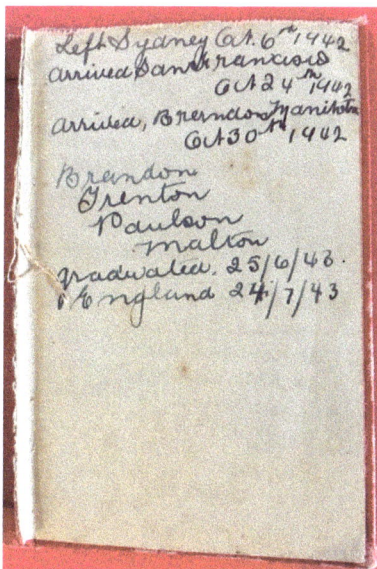

Helps to Worship: a manual for Holy Communion and Daily Prayer.

CHAPTER 3
RAAF SANDGATE

Charlie's enlistment took place at the RAAF No. 3 Recruitment Centre in Brisbane, having undergone a second medical examination by another Medical Officer with a signature scrawl, also illegible. There was a long list of medical conditions and physical defects, which did not present in Charlie. The Medical Examiner also declared Charlie could see the required distance with either eye, his heart and lungs were healthy, he had the free use of his joints and limbs and he wasn't subject to fits of any description.

Again, according to the Air Force Regulations, Charlie took the oath to serve King and country, with the additional clause, *"for the term of the duration of the war and twelve months thereafter, or until sooner lawfully discharged, dismissed or removed"*.

Approved by the Attesting Officer, P. Forman P/O (Pilot Officer), whose signature was stylish and legible, Charlie was appointed to No. 3 ITS (Initial Training) School based at RAAF Station Sandgate.

His RAAF life would finally get under way with his formal blue uniform and overalls for everyday work, plus a little blue book. Of similar size to the red book, The New Testament, which had been Presented by the Queensland Auxiliary of the British and Foreign Bible Society was to provide comfort and guidance for what was to come. The front page featured the emblem of the King's crown and below it his message.

> A Message From His Majesty The King –
> "To all serving in my forces by sea, or land, or in the air, and indeed, to all my people engaged in the defence of the Realm, I command the reading of this book.

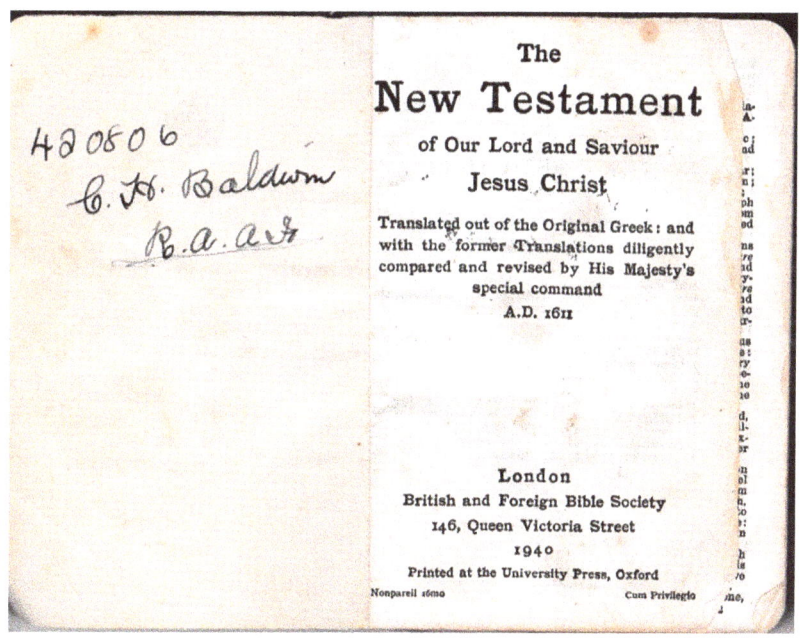

For centuries the Bible has been a wholesome and strengthening influence in our national life, and it behoves using these momentous days to turn with renewed faith to this Divine source of comfort and inspiration."

15 September 1939

In the front of the small bible Annie had written Charlie's name and RAAF Service number, 420806, but there were no entries to record dates or places after Charlie's arrival in England.

RAAF Station Sandgate, was located at Brighton, an outlying coastal suburb of Brisbane. It had become operational in July 1941 and an opening ball was held at Sandgate Town Hall to mark the event. An RAAF rifle range and an observation post were placed on a nearby farm. Training began with a few weeks learning the basics of military life, marching, bayonet and rifle drill plus aircraft identification, progressing to lessons in mathematics, navigation and aerodynamics.

Out of the initial trainee class of 1941, twenty graduated as RAAF aircrew. Of this number only three survived World War II. This told the story of the odds on returning home again, with the odds lessened for the aircrew who flew with Bomber Command.

Charlie would always remember the day of his arrival at Sandgate, 7th December 1941, the day the Japanese attacked Pearl Harbour.

He recalled a commotion at the camp gate that night when a guard was found unconscious.

> *Three rounds were fired in the air and for the next two hours everybody "galloped about", thinking the Japanese had landed there. Later it was discovered the guard had been hit by a brick thrown over the fence.*

CHAPTER 4

NARROMINE, NSW 1942

There was no mention of leave for Christmas 1941, having just started training with Air Crew 2 at Sandgate. On 16 February 1942, Charlie was re-mustered to Air Crew V (P) and promoted to Leading Aircraftman (LAC) on 25 April. After passing the initial training No. 26 Pilot's course, he departed Sandgate on 23 June for his posting to No. 5 Elementary Flying Training School (EFTS) at Narromine, NSW, a small town 400 kilometres inland of Sydney.

This RAAF pilot training unit had been established in June 1940 as part of Australia's contribution to the Empire Air Training Scheme (EATS), also known as the British Commonwealth Air Training Plan (BCATP). Both were referred to as simply, "The Plan", which began in 1939 as a joint aircrew training program involving the United Kingdom, Canada, Australia and New Zealand. Conditions for aircrew training were better located far from the UK and the fighting in Europe. Canada and Australia were ideal, offering wide expanses of land away from populated cities. Another joint training program operated in South Africa with the South Africa Air Force (SAAF).

CHARLIE – a journey from the Aussie bush to battle skies far away

Above: Charlie LAC (Leading Aircraftman) Right: Charlie in RAAF uniform

**PER ARDUA – AD ASTRA
Through Adversity
to the Stars.**

Of the Australian EATS graduates, more than 10,000 served in RAF Bomber Command's strategic air offensive against Germany and Italy. Of these, 3486 lost their lives in skies over Europe with 610 killed during training under Bomber Command – more than half of the total of all RAAF personnel killed in action, and almost 20 per cent of all Australian combat deaths in World War II.

The Narromine Aero Club's airfield, requisitioned by the RAAF buzzed with pilot training. The EFTS provided a twelve-week introductory flying course to personnel who had graduated from one of the RAAF's Initial Training Schools. There were two stages in flight training, which was on a simple trainer, such as a Tiger Moth. The first stage involved four weeks of instruction, including ten hours of flying to determine trainees' suitability to become pilots. Those who passed this grading process then received a further eight weeks of training, including sixty-five hours of flying at the EFTS.

Pilots who successfully completed this course were posted to a Service Flying Training School (SFTS) in either Australia or Canada for the next stage of their instruction as military aviators. Others moved on to different specialties, such as Wireless Schools, Air Observer Schools or Bombing and Gunnery Schools, all operating according to RAF training.

At each stage of training and testing to continue with pilot training as part of the EATS scheme, meant disappointment for those being designated to other airmen roles. Some had set their heart on becoming a pilot, perhaps being influenced by childhood heroes.

The chances of being selected for pilot training were likely to favour those who had attained secondary or tertiary level of education before being employed as civil servants, bankers, teachers and clerks in various fields of work. Some had already undertaken flying lessons before enlisting and being a sportsman added to the all-round character that made the cut.

Few recruits were farmers or manual workers, although it was noted that anyone who rode a horse had a good sense of balance, a handy attribute in flying a plane. Some of them, who perhaps had no choice in the job they were doing, realised that being accepted into the programme could lead to better prospects in the future. Those who had mechanical knowledge could become a flight engineer or work as ground crew.

Trainee pilots at No 5 EFTS Narrowmine, NSW. Charlie, standing far right.

Charlie with his sisters, Ena on the left, holding baby Wendy and Jacqueline (Johnny) at Merton, the family farm at Innisplain.

Charlie was free to enlist, knowing his brother was working on the family farm to help Annie after their father died. Although life on the land was close to Charlie's heart, he understood the value in having a good education and applied himself diligently, and later in life ensured he could provide his children with a good education.

Less than a week after Charlie arrived at Narromine, Annie recorded the date of her son's first flight. The course required ten hours of flying before the trainees took their first solo flight, a momentous occasion for Charlie, a young man from the bush with limited formal education. Annie would have been quietly proud, albeit with some concern for his safety.

At that time he was paid as a Temporary Sergeant. On 16 September, Charlie's twenty-third birthday, he completed the pilot elementary course and arrived home on a week's leave before he left for No. 2 Embarkation Depot at Bradfield Park, Sydney. During this brief time at home, Charlie would have stored precious memories of Merton and the family to keep them close to his heart and boost his spirits in the days ahead.

He also travelled to Nambour in SE Qld to see his sister Bonnie, where she was living after she'd trained as a nurse, married and had a baby girl, Wendy. Sadly, her husband, Brabazon Casement (2/9th Infantry Battalion) had been killed in the Middle East when Wendy was only a few months old. Bonnie continued nursing at Nambour, while her mother and siblings cared for Wendy at Merton.

Annie understood the consequences of war, with Bonnie's recent bereavement and during WWI two of Annie's brothers had served with the AIF. Thomas had been severely wounded and John spent 20 months as a POW (Prisoner of War) in Germany. So it would be with an anxious heart she farewelled Charlie, who had always wanted to fly.

CHAPTER 5

Voyage Across the Pacific 1942

Embarkation from Sydney took place on 26th September 1942 on the Dutch ship, *MS Noordam II*. A sea voyage would have been a new adventure for Charlie. In his recollections, he makes no mention of suffering sea-sickness, but perhaps rough weather would remind him of a bucking horse. Disembarking 18 days later at San Francisco, his journey continued by train through beautiful scenery along the West coast of America to Vancouver. For Charlie, he'd written that it felt like being on a holiday.

At home, Charlie's brother, Elwyn wrote a letter to their sister Bonnie on 2nd Nov 1942, regaling his fun at various local dances, with at least one every weekend. It wasn't all work at home on the farm and no play for this larrikin lad of 20. With Bonnie away nursing, Elwyn assured her everybody was ok at Merton, especially little Wendy, who was '*105% good all the time and in all the mischief imaginable*'. He also gave Bonnie an update on Charlie –

> *We had a cable from Chas. He landed safely. Dunno where but I suppose Canada.*

CHARLIE – a journey from the Aussie bush to battle skies far away

PER ARDUA AD ASTRA

MS Noordam II, which had been operating as a passenger-cargo service by Holland America Line (HAL). Her maiden trans-Atlantic voyage in 1938 was from Rotterdam to New York, but in 1942 she was converted to a troopship suitable to transport up to 2,400 troops. Throughout the war Holland America continued to manage the ship with a Dutch crew as well one or two American Officers, who would be in charge of the military on board. During the war, the Noordam transported 70,000 servicemen.

On arrival in Canada, Charlie was attached to the Royal Canadian Air Force (RCAF) and training began at the Manning ("M") Depot, Brandon in the Province of Manitoba. After a month of daily physical education, marching, rifle drill, foot drill, saluting and other routines, learning to bathe, shave, shine boots, polish buttons, maintain their uniform and behave in the required manner, a classification test and selection committee would then decide where the trainee would be placed, with aircrew or ground crew.

As a graduate of the No. 5 Elementary Flying Training School in Australia, Charlie had been selected for the next stage, 16 weeks at a Service Flying Training School. The Service schools were military establishments run by the RCAF or the RAF.

There were two different types of SFTS, one for the fighter pilot stream and the other for trainees in the bomber, coastal or transport pilot stream. For the first 8 weeks the trainee was part of an intermediate training squadron; for the next 6 weeks an advanced training squadron and for the final 2 weeks training was conducted at a Bombing & Gunnery School.

By late 1943, more than 130,000 personnel from Great Britain, Canada, Australia and New Zealand had graduated from 107 training schools across Canada. Over 100,000 administrative personnel were involved with operating the training schools and an additional 184 support units at more than 200 locations across Canada.

CHAPTER 6

Canada November 1942

Before arriving at No.12 Service Flying Training School (SFTS) at Brandon, Manitoba, Charlie enjoyed leave in Winnipeg, where he stayed at the Marlborough Hotel. He used the hotel's letter paper featuring its logo to write to his sister, Bonnie, dating the letter Sunday 8 November 1942.

Charlie's letter to Bonnie had been posted to Mrs Bon Casement on 11 November 1942 from Brandon, just three days before she re-married. The letter had passed censorship, stamped CENSOR 121. It had arrived in Australia and been forwarded to Bon, now Mrs V. Eglington.

Dear Bon,

I sent you a few postcards a few days ago so you will probably get these together. I've been spending a few days leave here and having a good time but it is so dreadfully cold. Yesterday was about 8 deg and what it's going to be like when it gets to 40 deg below as it does, I don't know.

Yesterday we went to see a football match but they can't play football here. It is only a game of throwing themselves on one another in the snow.

On Friday we went out to a park and saw some polar bears, wolves, coyotes, moose, reindeer, elk, buffalo and several varieties of rabbits which were very pretty. On our way out we went over a bridge over the Assiniboine River which was almost frozen over. There were places where the water was still running underneath the ice. I took some photos out there. I got in the cage to take one of the wolves. I haven't had them developed yet. I'll send some home in a while and I'll get them to send them on to you.

I heard some news on the wireless today that was put on especially for airmen serving in Canada. It was all about the widespread rains in Australia, the favourites for the Melbourne Cup and several other things.

The war news in North Africa is very encouraging now and it look as though they are going to put Italy out of it altogether.

Remember me to the Eglingtons, will you. Give my love to the darling Wendy when you see her. If all goes well I'll be leaving this country after winter which I hope to do and most likely go to England. Well Bon, I can't think of anything else at present so will say goodbye
with love
from Charlie xxx

Training began, but during this time Charlie succumbed to the snowy weather and spent six weeks in hospital recovering from pneumonia. At Brandon, pilot training was carried out on the Cessna Crane, a twin-engined advanced trainer aircraft designed and made in the USA and the Avro Anson, a British twin-engined, multi-role aircraft.

Then in January 1943 he was posted to RCAF base at Trenton, located close to the northern shores of Lake Ontario. It was the largest training centre of the British Commonwealth Air Training Plan during WWII, hosting the RCAF Central Flying School, No.1 Air Navigation School (until 1942), No.1 Flying Instructor School, and No.1 Composite Training School.

Having completed 24 dual and 24 solo flying hours, Charlie was "scrubbed out", or "washed out" as the Canadians termed it, and taken off pilot training to be reclassified as an air bomber

(A/B), also referred to as the bomb-aimer. He wasn't the only one, as 20 others were also "scrubbed out" at the time. It didn't always mean that those designated to other roles weren't good enough to be pilots, as numbers had to be adjusted with changing circumstances. It would have been disappointing to have made it through that far and then assigned to a different role, but Charlie would have applied himself equally to this speciality assigned to him.

The next training was at No.7 Bombing and Gunnery School (BAGS) located at Paulson, Manitoba, where the Australian and New Zealand airmen commemorated Anzac Day. Paulson was located near the large Dauphin Lake, which was ideal for bombing practice. Dozens of airfields were constructed in specific locations across the country, seemingly random, but with an eye to the post-war years when the airfields would be turned over to the local communities.

ANZAC Day 1943 in Canada was commemorated by Australian and New Zealand airmen who were stationed at Paulson in Manotoba.

CHAPTER 7

Land of Snowy Winters 1942- 43

More leave for Charlie and more news to send home, before his posting to No 1. Air Observers School (AOS) in Malton, Ontario [now the Toronto Pearson International Airport].

Air Observers were later referred to as Navigators and recruits undertook 8 weeks of training in an Avro Anson aircraft. The techniques were both visual pilotage and dead reckoning, using an aeronautical chart, magnetic compass, watch, trip log, pencil, Douglas protractor and Dalton navigational computer, a circular slide rule that was used in flight planning, estimating consumption of fuel, wind adjustment and time en route to the target. In flight it could also be used to calculate ground speed and how the wind was affecting the course set for reaching the bombing target. It was nicknamed "the whizz wheel".

As well as a portrait style photo of himself in uniform, Charlie sent Annie a postcard from Banff, in Alberta. From this land of snowy winters, Charlie would take home his own souvenir, a permanent reminder of his new cold climate experiences. A scar under his chin was the result of an encounter with an opponent's hockey stick during an ice hockey match in Winnipeg.

The Banff Springs Hotel, situated in the Banff National Park, offered Single room daily rates between $6.50 and $10.00 depending on the view of Sulphur Mountain or Bow Valley. In addition to the entertainment features and sports attractions the mountain resort was nestled in a setting that was home to buffalo, elk, mountain goat and other wild animals.

By mid-May, when Charlie arrived at Malton, the season had changed. He flew several navigation and aerial photography runs

To Mum, Love from Charlie, Paulson, Canada 1 May 1943 On the back of the photo: Banff
The land of:-Eating, Sleeping, Skating, Skiing and Keeping company with some of the prettiest girls in the Continent

along the coast of Lake Ontario and down towards Niagara Falls. The scenery made quite an impression on Charlie, describing it as:

> *The prettiest sight I'd seen in my whole life. We were up about 3,000 feet flying over the apple and peach orchards, and they were all flowering. It was pink and white for miles. It was a tremendous sight. The air was clear and still and the weather was so nice. It was great doing that for a while.*

On 25 June 1943, Charlie was awarded the Air Observer's Badge, graduated as Air Bomber and was appointed to a Commission in the General Duties of the Canadian Air Force.

To Mum, Love from Charlie Paulson, Canada 01/05/43

Charlie awarded The Air Observer Badge 25 June 1943.

Before being shipped to England, he went to New York on leave. The young Australians, in their blue uniforms stood out wherever they went. The Americans didn't always understand the Aussie accents, but were very friendly and the Canadians were wonderful, Charlie remembered –

> *A chap I met who worked for the Globe newspaper showed me around the place and I had a nice time down there. I went to the Stage Door Canteen which was a famous club for Service people, organised by film stars and celebrities and that was marvellous. A Canadian soldier had also given me his cousin's address in New York and we went around a bit together too.*

From 15 July 1943, on departure from Canada at the Overseas Transit Depot, Halifax, Charlie became attached to the RAF. This second sea voyage was shorter than the first across the Pacific, but the conditions more difficult, as he recalled:

> *The German U-boats were bad at the time so it was a fast, unescorted crossing in the French ship, Louis Pasteur, in just over five days and with 6,000 servicemen on board, it was only possible to have two meals a day.*

CHARLIE – a journey from the Aussie bush to battle skies far away

The Louis Pasteur, a turbine steam ship built in 1939 was designed to carry 751 passengers, but with the outbreak of WWII, would be used extensively as a troop transport and military hospital ship between Canada, South Africa, South America and Australia, also helping with the movement of Prisoners of War. Altogether, it carried 220,000 troops, and 30,000 wounded, and travelled 370,669 miles during the war. After the fall of France to Germany, the British government took over the ship, placing it under Cunard-White Star management.

CHAPTER 8

Arrival UK July 1943

British shores must have been a welcome sight as they disembarked at Liverpool. From there Charlie was sent to No.11 (RAAF) Personnel Despatch and Reception Centre, which had been relocated from Bournemouth to Brighton in May 1943. RAAF servicemen were billeted at the Brighton Grand or Metropole Hotels until being posted for further training, with the wait often being three to four months.

PER ARDUA – AD ASTRA

After a month's wait, Charlie was sent to No. 4 Observer Advanced Flying Unit based at RAF Station West Freugh, located in the Scottish area of Dumfries and Galloway. On a typed list of his Postings and Movements from enlistment to demobilisation, Charlie had written (Lysander) beside West Freugh, in a similar way to the place names he'd added beside the abbreviations for Operational Training Units. Instead, Lysander referred to a special little plane used for missions to ferry secret agents into and out of Nazi-held territory by night, as well as drop

The Westland Lysander was used primarily to drop off and pick up secret agents deep in enemy territory.

supplies critical for the Resistance, and retrieve allied airmen who had been shot down and were on the run. A reference to the Lysander, nicknamed Lizzie, being used in air gunnery training would be similar to Charlie's experience during his training at West Freugh, which served as a bombing trials unit with training facilities for observers, navigators, and bomb-aimers.

On 12 October, 1943, following six weeks navigation and gunnery training, Charlie was posted to No. 27 Operational Training Unit (OTU) at RAF Station Lichfield, Staffordshire, England. It was also known locally as Fradley Aerodrome, being situated in Fradley, two miles north of Lichfield and during WWII was the busiest airfield in Staffordshire. Its role was to form and train aircrew for front line bombing operations using Vickers Wellington bombers. The crews, largely from Australia and other Commonwealth countries, were then posted to their allocated squadrons, mostly in Lincolnshire.

Christmas Day 1943 presented Charlie with a promotion to Temporary Flight Sergeant, too late to include that news in the Christmas card already posted home to Bonnie and family. They would hear that news later, along with news of his move from Lichfield to No. 30 OTU at RAF Station Hixon, also located in Staffordshire. He'd taken the unusual step of invoking the provisions of an RAF regulation, which allowed a airman to withdraw from a crew if he could provide satisfactory reason. This caused a certain amount of "flak", as he put it.

Before leaving Lichfield to join an English crew at Hixon, Charlie learned that two fellow airmen had considered taking similar action, but reneged at the last minute. Later he recalled the issue as only involving Australians, but he didn't disclose further details or explain his reason. After his various postings in Canada and time spent before arriving at RAF Hixon, Charlie was keen to get on with training, so the incident or situation must have been serious for him to take this step.

On Australia Day, 26 January 1944, Charlie began training with a new crew at RAF Hixon, the airfield being 7.5 miles (12.1 km) east of Stafford, with railway lines along its northern and western boundaries, this proximity causing problems as a result of aircraft crashes. It was a very busy base, with added air traffic of American aircraft, as the United States Army Air Force (USAAF) base was close by at Stone. General Patton also alighted at RAF Hixon to visit the Prisoner of War camp located at Wolseley Road, Rugeley.

No. 30 OTU had been formed as part of RAF Bomber Command to train night bomber crews, which were often sent on missions over France and sometimes Germany to drop propaganda leaflets. Aircrews in training often had limited time to become familiar with each other and the complex skills required for combat flight before they were tested in action on these raids, with the codename, 'Nickel Raids'. These were just as dangerous as actual bombing runs as the enemy had no way of determining that they were only dropping leaflets.

The loss of crews and aircraft from the training units like those at Lichfield and Hixon was significant and a sombre awakening to the reality of war fought in the skies.

The twin-engined Vickers Wellington aircraft were used in training, but by the time Charlie arrived, they had done their duty and were experiencing mechanical failure due to age and daily use. Subsequently, accidents frequently occurred. After several crashes where the airframes had come to rest just beside or even on the railway tracks on at least two occasions, special phone lines were installed between the tower and the signalling staff controlling the railways in the event of an airframe fouling the railway lines.

**A Vickers Wellington aircraft at No. 30 OTU
(Operationa Traiing Unit) RAF Hixon, Staffordshire**

CHAPTER 9

An Accident 9 February 1944

Just two weeks after his arrival at RAF Hixon, Charlie would experience first-hand the common occurrence of accidents occurring during training. It wasn't unique to RAF Hixon, as indicated by graves close to many of the training locations, including 23 Aussie graves nearby Lichfield. Close to Australian training centres graves also told the story of lives lost before they left home soil to fight the war overseas.

On 10 February 1944, a twin-engined Wellington bomber began a practice navigational and bombing exercise, with two instructors and five aircrew on board. Charlie, under instruction was acting as bomb-aimer. This day would be embedded in his memory, as he clearly recalled:

> *Shortly after take-off when the aircraft had only got to about 150 feet, the port engine began malfunctioning, and very soon caught fire. Strong gusty winds made conditions very difficult.*

The pilot, a policeman in civilian life, had very little experience on these aircraft. He immediately turned back to the air-field.

Approaching the perimeter fence after completing the circuit, the wheels were down, but we were short by quite a margin. The aircraft hit the ground very hard, bouncing high, right over the top of two other Wellingtons in the parking bays. The aircraft crashed on the other side of the fully fuelled, stationary planes. It could have been a terrible inferno as we were still fully fuelled with 800 gallons on board. The plane went nose down into the ground in a relatively flat trajectory. Gusty wind spread flames quickly along the port side of the aircraft and along the top, burning off the highly flammable fabric covering the geodetic steel framework.

Before the aircraft crashed, I had lowered the perspex astrodome according to emergency drill procedure and then sat on the floor with my back against the main spar, facing the rear. I was alone in this section of the aircraft and couldn't immediately stand up without assistance from framework within the interior of the aircraft. When I stood up beneath the escape hatch through the astrodome, my chin was about level with the roof of the bare framework.

I then placed my forearms right up to the elbows out through the opening with my hands open in order to lift myself out. Not realising the level of heat in the framework, I put all my weight on my arms and hands with the instantaneous and complete effect of having all my skin from my elbows down the length of my arm folded up or hanging from my fingertips.

It was immediately obvious that whatever the circumstances dictated, nobody was going to make their escape through the astrodome. I knew I had no hope of getting out that way. The fire was burning the fabric covering completely off the plane and the framework was red hot.

I then thought of the rear turret but immediately realised that the rear gunner had wound the turret to a "side-on" position in order to escape that way, thereby rendering it impossible to enter the turret from within the aircraft.

As I was turning to look elsewhere, the navigator Fred Flattery emerged from the very thick, hot column of yellow smoke, heavily impregnated with myriads of burning sparks coming from the starboard wing roots. The starboard motor was burning.

Fred said to me, 'Charlie, we have to get out here. We can't get out up front. I tried it and it's not possible.'

I showed him my arms and said, 'We just have to get out up front. Just look at this.'

Things looked really crook and we were beginning to think there wasn't any way we could escape.

There was no panic and I saw things quite clearly. He immediately understood. We stood face to face without speaking in the few seconds before we proceeded forward.

Fred went first and I followed closely on his heels. The density and heat of the smoke forced us to close our eyes and my hands were a distinct liability. I suffered burns to my face as we stumbled through the plane. There were some unexplained objects and broken bits that had been flung around, obstructing the passageway and creating some difficulty. I was starting to feel dizzy from holding my breath.

When we cleared the smoke and flames about our feet, we were standing beside the pilot's seat. The windscreen had been knocked out earlier. We realised the hot metal of the open framework constituted a real danger if we attempted to walk across it and jump down, lest even one of our feet slipped through. In that case we would not be able to extricate ourselves from it.

Without any hesitation, Fred left first and I followed. We stood on the pilot's seat, folded our arms across our chests and flung ourselves forward at full length along the nose of the aircraft and rolled off onto the ground about 7ft below. I raced away about 10yds and collapsed on the ground. Perhaps only seconds later, I heard the most high pitched, agonising shriek from behind me.

I looked back and saw the radio operator, Gus (Sergeant Arthur Welstead) hanging by the leg straps of his parachute harness from the skeleton of the aircraft near the joining of the aircraft nose and fuselage, with his feet just above the ground. In terrible agony with his hands badly burnt, he couldn't release the harness.

I immediately got up and went back in an endeavour to get to him, but I collapsed about 6ft from him, immediately in front of the port motor, the cylinders of which were exploding spasmodically and showering molten metal fragments about.

Having collapsed I was unable to do anything for this wonderful young Englishman of 19 years. I could only crawl away from there with much difficulty. Gus was eventually released from the parachute by the village baker who lived near the airfield.

It was noted that after the aircraft hit the ground the fire was all over in a relatively short time, as witnessed by a small group of people gathered together, standing around for a while. All personnel on board were accounted for, with someone pointing out that Charlie's jacket was still smouldering. He hadn't been aware of it.

The Unit doctor came quickly, but could do little to give Gus comfort until placing him in the Unit Infirmary for the night. Fred and Charlie were also stretchered to the Infirmary. Late in the evening, when it would be expected the patients were asleep, Charlie heard two nurses come in. He would always remember the anguish he felt lying there, as they quietly wheeled Gus out.

> *I was sorely tempted to say goodbye to him openly, but was afraid of upsetting the nurses and remained silent. It was a heart-rending time.*

Later, the local village baker, Cyril Fradley was awarded the British Empire Medal (Civil Division) for his actions, as published in The London Gazette. Established in 1665, the Gazette was the official record of the Crown, publishing announcements of royal births, ceremonies, awards and accreditations for the public records.

"The KING has been graciously pleased to give orders for the undermentioned award of the British Empire Medal, and for the publication in the London Gazette of the name of the person specially shown below as having received an expression of Commendation for his brave conduct.

> *CENTRAL CHANCERY OF THE ORDERS OF KNIGHTHOOD, St. James's Palace, St. James's Palace, S.W.i.30th May, 1944.*
>
> *Awarded the British Empire Medal (Civil Division):—Cyril Fradley, Baker and Confectioner, Hixon, Staffs.*

When an aircraft crashed and caught fire three members of the crew jumped to safety just as the flames were spreading from the port engine to the pilot's cabin. The fourth member of the crew was caught by his parachute harness as he was climbing out of the cabin and was left suspended on the side of the burning fuselage. Fradley, and others, ran forward to release the airman but they were driven back by the intense heat. Fradley, however, made a second attempt. Undeterred by the flames and exploding ammunition, he returned to the blazing wreckage, grasped the airman round the legs and lifted him clear. Unfortunately, the airman died some hours later from his injuries. Fradley was severely burned about the face and hands when making his gallant attempt to save life."

Numb. 36535   2487

THIRD SUPPLEMENT
TO
The London Gazette
Of FRIDAY, the 26th of MAY, 1944
Published by Authority

Registered as a newspaper

TUESDAY, 30 MAY, 1944

CHAPTER 10

Convalescence February - June 1944

After a night at the Hixon Unit Infirmary, Charlie was taken to Stafford General Infirmary, listed as seriously ill and a few days later was transferred to a large RAF hospital at Cosford near Wolverhampton, where he was placed in a ward for burns patients. The eminent New Zealand surgeon, Dr McIndoe visited the hospital regularly and on one occasion briefly examined Charlie, who was in a saline bath.

The ward contained a number of very serious cases and it was impossible for Charlie not to be aware of the nature and extent of the sheer agony and distress they were suffering, knowing the debilitating disfigurement they would bear thereafter. Not all survived their injuries.

Charlie felt his injuries were of a relatively minor nature compared to others less fortunate than himself. He had kept his flying helmet on, thereby protecting his ears, but his blackened face took some time to return to normal.

CHARLIE – a journey from the Aussie bush to battle skies far away

After his discharge from hospital, Charlie, a Flight Sergeant by this time, was convalescing on leave, when a family invited him to stay with them in London. The husband, who was Australian had stayed on in England with his English wife after WWI. They suggested a visit to Devon where Charlie was a guest at Dartington Hall.

The estate was owned by a rich Englishman, Leonard Elmhirst and his wealthy American wife, Dorothy Payne Whitney, who had family connections to the Packard Motor family in Detroit. Elmhirst, an agronomist and philanthropist, together with his wife bought the property in 1925, restored the historic medieval Hall and in 1935 set up a registered charity, The Dartington Hall Trust. The estate comprised various schools, colleges and charitable organisations, which Charlie thought was a remarkable place with dairy farms, quarries and a music school.

Dartington Hall, Devon

Also staying at the estate was a Canadian Padre and two British soldiers. The Padre considered Charlie could benefit from more recovery time and offered to arrange extra leave, which Charlie appreciated, as he was still "a bit shaky on it". It had taken weeks to walk unaided and would take a while longer to regain the same measure of mobility he'd previously had. In particular his right shoulder suffered from the impact of hitting the ground as he escaped the burning plane.

The padre knew the group Captain of No.10 Sunderland Flying Boat Squadron and organised a rail warrant for Charlie to visit the RAF station at Mount Batten near Plymouth.

Australia had purchased the Short Sunderland aircraft before the war and a small group of squadron personnel had been sent to the UK for training with the plan to fly them back to Australia. Before they could return home, war was declared and the Australian government ordered the squadron to remain there to assist in the British war effort.

After this two extra weeks of leave, Charlie returned to RAF Hixon, although the crew involved in the accident with him, had been split up and once recovered, were reassigned. The day of his return, four months after the crash, was 6 June 1944, the date known as D-Day. On that bleak grey day, Charlie was walking back to the Station when he saw some American Flying Fortresses heading west, recalling, 'I knew there was something on'.

CHAPTER 11

Farewell Fred

On returning to RAF Hixon, Charlie was placed in the unit hospital for observation, before being declared ready to resume training. Restless at night, he was sleeping fitfully when an incident happened that would impact him quite badly. It would be another memory that stayed with him:

> *I was dozing when I was disturbed by a brilliant light in the large window on the opposite side of the ward. I hurried across the ward in real fright as I realised it was something that shouldn't be, with the sky ablaze. To my horror, I saw an aircraft, which I knew to be a Wellington engaged on night exercises, on fire from one end to the other. It was flying at about 1,000ft, exploding with various portions of it falling to the ground.*

In that instant, Charlie's perception of reality became distorted for a few seconds, taking him back to the crash he had been in.

Other patients were asking me what had happened, but I was unable to speak to them. Later in the evening, the aircrew were brought into the unit hospital, obviously in great agony. Their cries of anguish affected everybody within hearing. None survived.

My great problem was that for some time I couldn't determine clearly whether I was dreaming or whether it actually happened. It was a strange twist of fate that it should happen just two nights after I got back to the station. I thought I was going insane, but I put up with it, then finally got over it.

However, that incident would continue to torture Charlie in nightmares – upsetting and unforgettable.

Shortly after this traumatic event, Charlie was pleased to see Fred again, when he turned up at Hixon, still suffering from his burns and being 'a bit knocked around'. He'd fractured his skull when he hit the concrete after propelling himself off the plane.

With nothing to do until resuming duty, Fred and Charlie each bought a bike and spent their afternoons cycling through the nearby countryside, down lanes bordered with hawthorn hedges. Charlie would have been interested in the crops and fields, sprinkled with wildflowers, poppies and cornflowers painting a picture so different to the countryside at Merton.

On one occasion they noticed a dance in a local hall and decided to join in, but were refused entry at the door.

> *We both considered this an affront, particularly as the hall was full of Italian prisoners-of-war, who were farm hands in the surrounding area. We objected strongly. Then Fred and I were joined by a dozen other British and Allied Servicemen in uniform, so we said to them, 'what about it, you fellas?'*
>
> *There was a police station across the road, and what must have been a dozen police lined up on the stairs watching the proceedings. They were looking at us and we looked at them, but I couldn't imagine them treating us badly. If it hadn't been for the fact the dance was almost over, there could have been a stoush.*

Fred and Charlie never flew together again, but had become great mates and kept in touch. Coincidentally they had both been training in Canada at the same time. While Fred was stationed at No. 4 Air Observer School at London, Ontario between March and July 1943, Charlie was located about 100 miles away at No. 1 Air Observers School at Malton. Afterwards when they met at RAF Hixon, it was likely their friendship began with the shared experience of training and living in Canada. Then escaping together and surviving the fiery crash at Hixon had forged a special bond. After recovery they were sent to different locations but kept in contact.

Towards the end of the war after a few months with no word from Fred, Charlie went looking for him at his family's home in Birmingham. Charlie learned the sad news that Fred had been killed in a collision between two aircraft over the mouth of the Somme River in France.

Fred had joined 460 Squadron, an Australian squadron stationed at Binbrook, Lincolnshire, only 20 miles from where Charlie would later fly with RAF 550 Squadron at North Killingholme.

Unfortunately, the 460 Squadron's motto, "strike and return" couldn't always bring them home. This squadron suffered higher casualties than any other unit of Australian army, navy or airforce during WWII.

RAAF Squadron 460 Binbrook, Lincs

A further search for Fred in records provided more information about his final flight. On 2 February 1945, Fred, aged 21, was the navigator on board the aircraft piloted by Group Captain KRJ Parsons DFC, the only member of the crew to survive the collision over the Somme. He and the wireless operator Flt/Sgt Cunningham were RAAF members and the other five aircrew all with the RAF.

Frederick John Flattery was the son of Frederick Sidney and Mary Magdalene Flattery, residents of Billesley, Birmingham. Fred's grave is at the Military Cemetery, Abbeville Communal Cemetery extension, in the city of Abbeville, on the Somme River in the Picardie Region.

CHAPTER 12

Return to Flying 1944

On his return to training, Charlie requested he might start flying firstly with Chief Flying Instructor, Squadron Leader Swan. His request was granted and he also flew with other experienced operational pilots. On a test flight with Australian Pilot Officer (P/O) Jock Cameron an engine failed, causing considerable anxiety but they landed without much difficulty.

The time came when Charlie was assigned to an aircrew captained by Flying Officer (F/O) Ted (Theodore) Gorak, an American in the Royal Canadian Airforce (RCAF). Three of the crew were Canadians: Bob (Robert) Mills (Navigator), Don Morrison (Mid-Upper Gunner) and Harold Graham (Rear Gunner). The Flight Engineer, Bob Whitehead was English and completing the crew of seven were two Australians, Stan Willis (Wireless Operator) and himself, Charlie Baldwin (Bomb Aimer).

They proceeded to No.1656 Heavy Conversion Unit (HCU) at RAF Lindholme, Yorkshire, where they would receive training in operating heavy bombers before being posted to an operational squadron. It involved more complicated devices and specific instructions on route, speed and operating within the concentrated stream of bombers. As part of the training, some of the HCUs were involved in bombing operations over Germany.

On 26 August 1944, the crew were posted to 103 Squadron at RAF Elsham Wolds, Lincolnshire, their arrival memorable by not getting off to a good start, as Charlie noted:

> *Immediately upon our arrival the crew was paraded before the (CO) Commanding Officer, who was also Wing Commander and for some reason was in a particularly bad frame of mind. Unfortunately for our pilot, Ted, as he did not have the insignia (wings) of his position in the crew attached to his battle-dress jacket. This was the outlet for a considerable volume of anger and frustration, obviously a result of something that happened at the station before our arrival.*
>
> *It was normal practice for a crew arriving at a squadron to be given six days leave before going on operations, but we were immediately ordered out on a gunnery practice over the English Channel, dropping ultramarine smoke markers on the water.*
>
> *Both gunners and I were to use them to shoot at. We hadn't been out long before being signalled to return to base, but there were a few seconds of hesitation before a response from Ted. On our return, the whole crew had to*

front the CO (Commanding Officer) for the second time within an hour. The atmosphere was tense after the CO tore into Ted on parade, but it became positively electric again, this time for his slow reaction to a signal.

Ted was subsequently ordered to join a night mission with an experienced crew, flying as the second pilot, referred to as "Second Dickie". This was the usual practice for the purpose of pilots' familiarisation before taking their own crews on ops. The raid to Koenigsberg was a long and dangerous assignment and the plane was shot down. The crew failed to return and we were then without a pilot.

This poor fellow, posted to and officially a member of the squadron the same as everybody else, never enjoyed the pleasure of sleeping here once, killed in action the first night. This was very unsettling for the rest of us.

Thrown into action immediately on their arrival, Ted's crew had no time to consider the dangers of what lay ahead, until their reflection on how quickly he was gone. Fortune was with two others in their crew that night, otherwise the crew could have been just four left, instead of a "headless" crew of six, as it was referred to. Charlie remembered:

As it happened, there were two crews in the squadron short of one member each, with two of our crew, Stan and Don required to fill the vacancies in two different aircraft to provide the squadron's full complement of aircraft for a raid that night. They both returned and the six of us were certainly ready for our six days leave. The next move for us was to have another pilot assigned to our crew as without a pilot we were not going anywhere and therefore quite unproductive.

Flying Officer (F/O) George G. H. Cowper, an experienced de Havilland Mosquito Pilot was put in charge of our now complete crew and we went back to Lancaster Finishing School at RAF Lindholme. George Cowper had never flown Lancasters previously and the crew had to operate "as one" on these aircraft and get to know each other.

Left to right: F/O Cowper, F/O Mills, F/SGT Baldwin, SGT Whitehead, F/SGT WIllis, SGT Morrison, SGT Graham

CHAPTER 13

The Lancaster

Each aircrew of seven men would become a close-knit unit, depending on each other when they encountered problems to carry out the mission successfully and survive, with the odds stacked against them. Each of them had specialised knowledge to rely on during a raid, helping to guide the pilot in determining their next course of action, as they dealt with the situation. They also needed to step into each other's job if one of them was injured or killed in flight. Having graduated from Pilot School before completing navigational observation, bombing and gunnery training, Charlie could act as the reserve pilot in an emergency, the usual secondary role for the bomb-aimer.

During flight, the role of the bomb-aimer was to man the nose-mounted gun turret and provide assistance to the navigator before crawling into the bomb-aimer's compartment in the lower section of the aircraft nose. Lying flat, he operated the bomb

CHARLIE – a journey from the Aussie bush to battle skies far away

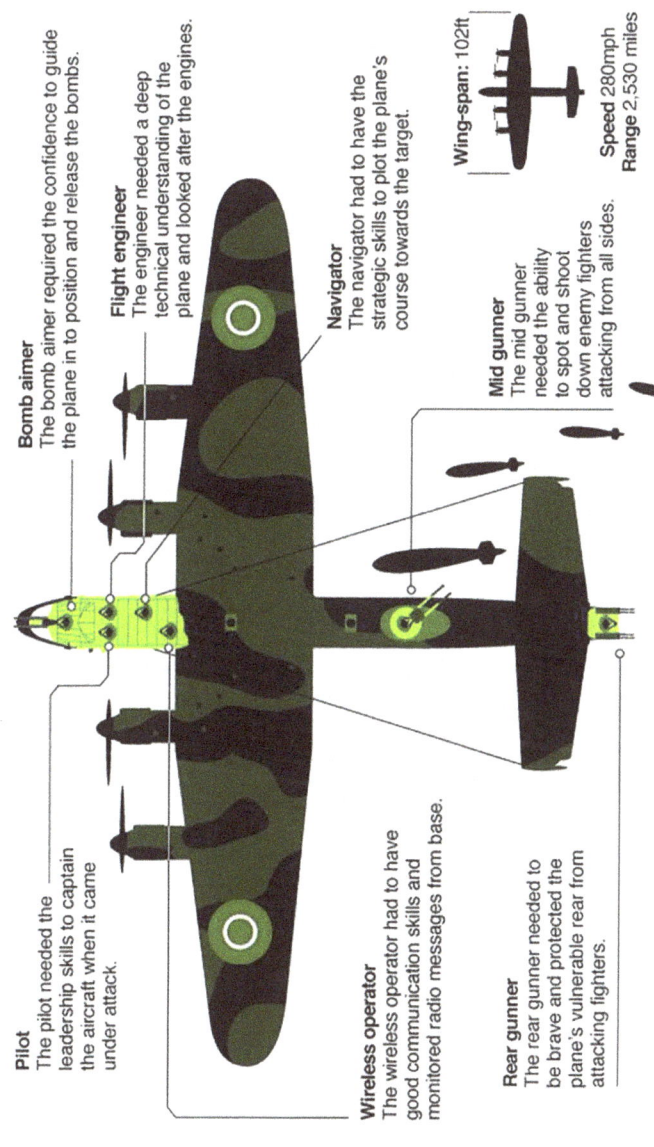

sight instructing the pilot on minute adjustments to the plane's course in order to obtain the best possible accuracy before releasing the bomb load.

Dropping the bombs was the most dangerous part of a mission, requiring the plane to fly level and straight. Markers would have been dropped by crews and the final "go-ahead" issued from the Bombing Commander, before the button was pressed to drop the bombs, activating a flash photograph to record the result. This proof of mission accomplished was counted towards the crew's tally of 'ops' necessary to complete a tour of 30. No matter what the reason, an aborted mission wouldn't be counted.

In pages of hand-written notes, Charlie had described details of the Lancaster and its operation. He'd included the width, length, power, speed, weight, fuel capacity and consumption. Then there was the armour, the type of guns in the turrets, the bomb bay, the bomb stations and bombs. From the bombs being loaded the crew could gauge the possible type of target and likely location. Charlie mentioned that the bomb load for their crew was predominantly a four thousand pounder with the balance being incendiaries. Navigational aids were of particular interest to Charlie, explaining the different systems used:

> *GEE: radar set receiving signal from three transmitters in Britain. Operated by the navigator they were very accurate but reception affected by enemy jamming as aircraft progressed and distanced from base.*

CHARLIE – a journey from the Aussie bush to battle skies far away

Lancaster Aircraft

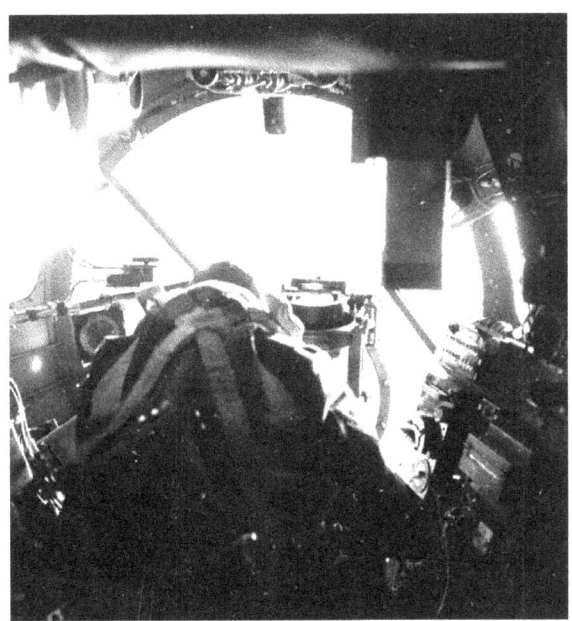

Bomb-aimer in position during flight

H2s set: self contained within aircraft and exceptionally good where shorelines of sea, lakes and somewhat to a lesser extent, mountains. In our case, I as bomb-aimer, mostly operated this set.

Pool: another receiver which was used to detect enemy aircraft travelling in bomber stream simply by the relative speed factor: easily identifiable where fighter aircraft were travelling at their normal speed, but where fighters slowed down to about bombers speed no identification possible that way.

Astra navigation: I usually used the sextant when not otherwise engaged and was of course quite familiar with these heavenly bodies suitably placed throughout the universe, on which to work when the weather permitted. The old type of sextant we used was hung (while being used) on a clip in the astrodome about "midship". It was operated by a gun trigger and had to be kept level with a bubble correctly positioned and maintained in the centre for a full minute while the trigger was operating. This could be difficult in rough weather.

A system operated whereby a small percentage of the total aircraft in an op, considered to have the best navigators were used by the base to assess the navigational findings and determine what the most accurate readings were and then transmit the result to all aircraft.

After we had done about three trips, I remember our navigator Bob Mills asking a seasoned navigator how often he took a navigational fix, and when he was told usually every 15-20 minutes, Bob said he was in a bit of a flap as he felt he should and was trying to take a fix every three minutes.

So I don't recall any problems of a navigational kind. This was an example of certain things one had to learn from first hand experience, and there were plenty of them.

CHAPTER 14

Finishing School

It was during our two weeks stay at RAF Lindholme that I had another hair-raising experience which shook me badly. In the morning of that particular day, there had been a mid-air collision between two American aircraft. A motley assortment of debris came floating down with everyone scrambling to dodge it, which was quite unsettling.

Later I was walking to lunch by myself when I saw a Lancaster flying across from my left to right at about 300yds. On this clearest of sunny summer days, a Hurricane fighter aircraft took off from the airfield on my right and struck the Lancaster and cut the rear turret right off, together with all control cables to the rudder and elevators, leaving the aircraft completely devoid of control.

At this point the Lancaster was heading about 35-40 degrees to the right of my direction and probably about 500yds distant. From that point the Hurricane crashed to the ground a little to my left. The pilot had parachuted but was killed on impact as the chute hadn't time to fully open due to the lack of height.

I was closely observing the Lancaster which was making a very long wide sweep to the right and losing height. No attempt was made to parachute. The aircraft continued on its circular path and soon appeared heading on its opposite original direction and over the top of the double-storey brick building, which housed permanent officers.

It was at about this point that I had to decide in which direction I should move to get away as far as possible from its path. The aircraft direction seemed to be getting harder to determine and I quickly decided to continue on my initial direction. I ran as hard as possible and flung myself into a ditch beside the road on the station boundary.

At about the time I landed in the ditch, the aircraft crashed into some buildings about 50 yds away, killing two ground staff and the remainder of the crew, the rear gunner having already been killed by the Hurricane on initial impact.

While this was happening, a ground staff man was riding his bicycle on the same road and in the same direction as me. He was knocked off his bicycle by the wing tip of the Lancaster just before it crashed. His physical injuries were minimal but he suffered severe shock, as I did too, especially upon seeing various parts of bodies of the victims.

By that stage I'd begun to feel demoralised and disconsolate to some extent. It had been suggested to me in a round-about and somewhat unofficial manner, that I might be thinking of going home, which was something that hadn't entered my mind. People were being killed daily, but whilst I had my share of misfortune up until then, I realised I was still well-off compared to many others and instantly dismissed the suggestion.

In fact I felt relieved when we arrived at 550 Squadron in North Killingholme, Lincolnshire.

Charlie's character would take a leaf out of his father's book, 'to keep a stiff upper lip and hang on'. His family's attitude and values were further upheld by faith and the teaching at the church's Sunday School. Many of the young men accepted for training in the RAAF shared this form of schooling, but may have been from very different circumstances. There was an emphasis on duty, service, steadfastness, loyalty and honesty, being important in the selecting a recruit with good all-round abilities; the physical attributes, skills and mental aptitude to be applied in what lay ahead.

For many of the airmen, it took a considerable period of time from the beginning of training until finally they were getting into the action. It included time spent waiting for a posting, or aircraft to be available or travelling in transit, the ocean crossings taking the longest amount of time. For Charlie, it was almost three years, including four months of recovery after the aircraft crash during training, before his first mission with the crew at 550 Squadron on 31 October 1944.

*CHARLIE – a journey from the Aussie
bush to battle skies far away*

Many Australian airmen had witnessed death and injury before flying their first operation over Europe, some never arriving in the UK. Graves beside the airfields in Australia and Canada were the legacy of accidents during training. The risk of death was ever-present due to mid-air crashes, poor visibility and lost bearings during training exercises, day and night. Always the weather conditions were crucial to the risks for aircraft and crew.

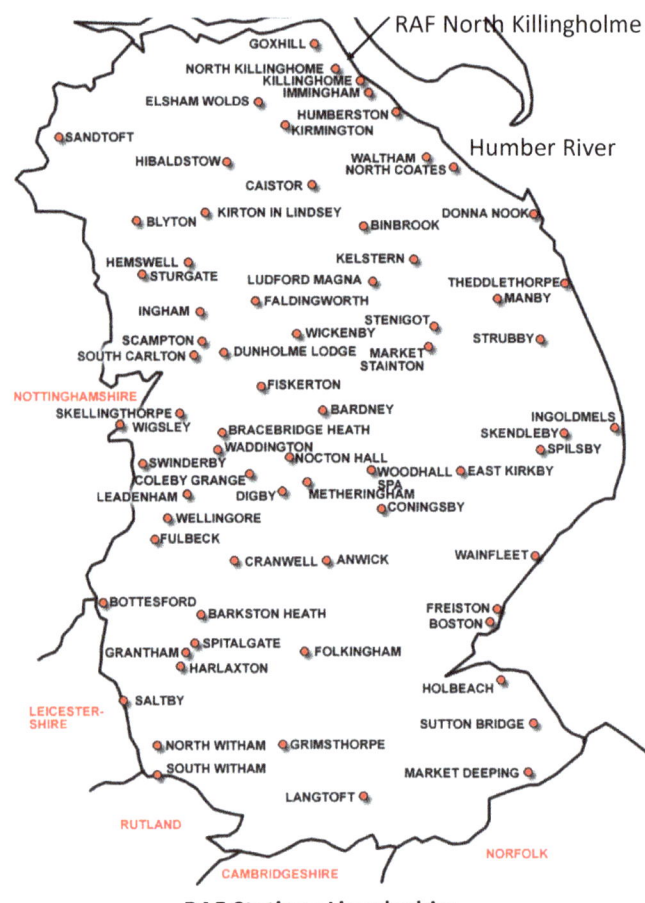

RAF Stations Lincolnshire

In the UK, it didn't help to be training on aircraft that could suffer mechanical problems, having already completed a lengthy workload of flying ops, with many aircraft being damaged and repaired. The RAAF suffered thousands of casualties in training units and flying schools.

CHARLIE – a journey from the Aussie bush to battle skies far away

Arrival at North Killingholme
Standing Left to right: Canadian Don Morrison (Upper Gunner), Australian StanWillis (Wireless Operator), English Bob Whitehead (Flight Engineer), English George Copwer (Pilot), Charlie (Bomb Aimer)
Front: Canadian Bob Mills (Navigator), Canadian Harold Graham (Rear Gunner)

CHAPTER 15

550 Squadron October 1944

550 Squadron was a heavy bomber squadron, formed on 25 November 1943 at RAF Waltham, located near Grimsby, a coastal seaport on the Humber Estuary in North East Lincolnshire. Fifteen miles to the north of Waltham, preparations were being ramped up for the squadron's move on 3 January '44 to Station RAF North Killingholme, which would be a sub-station of 13 Base, the headquarters being at Elsham Wolds.

Time was of the essence to get everything fully functioning for the aircrews to resume their part in No. 1 Group Bomber Command's attacks on Germany. A huge effort had been made in organising the set-up although the facilities were far from comfortable to start with, so it was a rather bleak beginning to the station located on a marshy area close to the Humber River.

By the time F/O Cowper and his crew arrived at North Killingholme on 16 October 1944, the station was much better equipped with an addition of extra crews, making 550 a three-

flight squadron. The squadron's aircraft were four-engined heavy bombers, Avro Lancasters, normally used in night operations to drop both incendiary and high-explosive bombs on enemy targets. In between operational sorties being carried out, there was a full training program including night-time exercises.

The 550 Squadron insignia, which featured a sword pointing upwards in front of flames and its motto, Per Ignem Vincimus "through fire we conquer" would be symbolic of the squadron's skill to force its way through the barrage of enemy fighter fire and flak. Charlie had already won through his first encounter with fire and was more than ready to get on with the task of fighting the enemy.

Daily life demanded a rigorous routine, the effects of which would test their physical and mental stamina. It was a pattern of limited sleep, focused preparation and tension as they waited to lift off, sometimes being "stood down" when an operational sortie had to be cancelled. Smoking had become more than a new social habit for Charlie, a common way for the aircrew to cope. As he said:

> *It was something to occupy the time during the interminable waiting, sometimes for hours, waiting for takeoff.*

For each Lancaster to be ready for every op, it took a combined effort by the seven aircrew, several ground flight mechanics and specialists to service the wireless and radar equipment. Several armourers were required to prepare, transport and load the bombs. The armourers worked hard, doubly so when ops were cancelled, having to disarm the aircraft. Drivers were needed for the tractor, petrol bowser, oil tanker and bomb "train".

The Flying Control Officer and assistants worked in plotting, navigational mapping and communications. Once the mission had been posted on the notice board, preliminary briefings were attended by pilots, navigators and bomb-aimers, with general briefings carried out afterwards. Holding the key factor was the meteorologist, since weather played a crucial role both at home and the target area. In all, it would require about 37 service personnel for each op carried out or stood down before take-off.

Women serving with the Women's Auxiliary Air Force, known as WAAFs, occupied a diverse range of work that included packing parachutes, roles in plotting, radar, intelligence, communication and as transport drivers. The Air Transport Auxiliary (ATA) provided civilian women pilots, who delivered aircraft and ferried planes between airfields.

Preparing for an op, the aircrew donned flying suits, jackets, boots, helmets and gloves, designed to protect them physically from the extreme conditions. The rear and mid-upper gunners were more exposed to the elements and often removed their fur-lined gloves, suffering frostbite as a result. Geared up, each airman took his parachute, an inflatable vest, nicknamed a "Mae West" in reference to the large-breasted US actress and the important escape kit to help his survival on the ground in enemy territory. Nothing personal, other than a lucky charm accompanied them on what could be their last journey.

Sticking to routines – having the same driver taking them to the same aircraft, the last minute cup of tea, a silent prayer or thoughts of loved ones and a nervous pee – meant entrusting their luck would hold out. The air crews would form a close bond with a particular aircraft, relying on both plane and its ground crew to bring them home. They would also run their own test or checks on the aircraft before takeoff. If the aircraft failed due to mechanical problems, an aborted flight could not be attributed to LMF (Lack of Moral Fibre).

The ground crew felt the loss of men and aircraft, working closely with them, waving them off, wishing them good luck, and were there to welcome them home if they were fortunate enough to return.

Raids would often be 10 to 12 hours long, with temperatures as low as minus 72° (F) at heights between 16,000 to 21,00 feet. The bombers were not pressurised like modern aircraft. Once airborne, every airman had to hope that his oxygen mask and supply functioned properly when it was switched on at 5,000 feet and did not suffer from icing in the frigid temperatures.

Standing beside a Lancaster: F/O Mills, SGT Whitehead, SGT Graham, F/O Cowper, F/SGT Willis, F/SGT Baldwin, SGT Morrison

The frequency of ops varied, but ever present were the perilous conditions and high odds of not returning. The dangers came, not only from the enemy fighters and damaging flak, but the long hours of intense concentration, very uncomfortable conditions on the aircraft, mechanical problems, adverse weather, disorientation and crashes before or after arriving back on home soil. Fatigue had to be overcome to maintain their mental agility in making calculations and decisions, so vital to their survival.

CHAPTER 16

Battle Skies

The skies above Europe were teeming with aircraft, Allied Bombers and Pathfinders trying to avoid enemy fighter planes. Bomber Command had been relentless in its campaign to defeat Germany from above, a concentrated effort by thousands of men with support from many more men and women on the ground. Squadrons would usually be assigned the task of dispatching 12-25 aircraft on a night operation and at least one of their crews would be expected to be lost every two night-time ops. On several nights, some squadrons would lose five or six of their crews in a single night, 42 crewmen from a single squadron who didn't come back from that one operation.

The first sortie for the crew captained by F/O George Cowper, was a raid to Cologne, the target being military objectives and communications. As one of 24 aircraft briefed to do the raid, records note Cowper's aircraft to fly at 19,000 ft. at 21.15. Aircraft didn't fly in formation, but as part of the bomber stream, which could depart in waves. Charlie described it as a rather quiet affair. All aircraft retuned safely.

Back-to-back ops were especially testing. At one stage, Charlie's crew made four trips in four nights which meant they could only sleep between 8am and noon. The afternoons were spent in briefings and preparation. Leaving England behind, all eyes would be straining into the night sky, hoping to see a night fighter approaching before it was able to open fire, or to see another bomber in time to take avoiding action before a collision occurred.

On a raid to Freiburg, Charlie remembered it was the first time they came up against the new German fighters. They could see the jets' vapour trails high above them.

> *They were quick to get down and we lost seven aircraft, although an Australian gunner shot down two of them. During this confrontation, I broke the golden rule, instead of keeping a sharp eye out for enemy fighters, I watched one of the Lancasters go down. The poor old "Lanc" was ablaze from one end to the other as I saw it crash.*

With fighters out in force, large numbers of parachutes could be seen, lit up by the fires below. After their attacks the crew would quickly head away from the target area to return home, endeavouring to avoid flak and fighter hot-spots. Vigilance had to be maintained or even increased at a time when tiredness could begin to affect them. The German fighters continued to hunt the bombers and slipped in behind them, slowing down their own speed to remain undetected. They tailed the bombers home across the English Channel to shoot them down over the sea or over the base before they could land. Charlie recalled this happening to a New Zealand crew, who were also with 550 Squadron and he noted that:

> *There were many acts of heroism, particularly on the part of bomber pilots who sacrificed their own lives getting the crew out after their planes had been disabled.*

CHAPTER 17

Christmas 1944

The weather, especially the adverse conditions of winter impacted their lives in the air and on the ground, when cloud, fog, icy conditions, or snow would prevent operations. Moonlight could also be crucial to the raids, affecting the visibility of markers and targets, but also making it easy to be spotted by the enemy. Strong winds would give navigators and bomb-aimers extra concern, as Charlie recalled on a raid over the Ruhr Valley, 'Happy Valley' as it was often referred to.

> *It was a battle to reach the target and then carry out the precise holding before releasing the bombs.*

Holding the aircraft steady for the photographic evidence after bombing, was made difficult by the sudden change in weight, with the added wind factor.

Due to thick fog and poor visibility, it was often necessary for returning aircraft to divert to other airfields, and may have caused

Charlie's crew to land at Witchford airfield on their return from a raid to Wanne-Eickel.

For ground staff and personnel at North Killingholme, Christmas Day 1944 was missing the cheer of the aircrew, with many of them absent. The aircrews who landed at Wendling were given a warm American welcome by the US 8th Air Force and undoubtedly a hearty Christmas meal. At North Killingholme, the menu included pea soup, roast turkey and stuffing, roast pork and apple sauce, potatoes roasted or creamed, Christmas pudding with rum sauce, jellies and mince pies, accompanied by beer and cigarettes.

The weather also affected life in general, with icicles hanging on buildings, snow drifts to wade through to get to the mess and runways to be cleared of snow, either by machine or hand-pushed brooms. Canada's cold had at least introduced Charlie to these conditions.

The new year, 1945 arrived, bringing severe cold with blizzard conditions. Icy roads, treacherous runways and gale-force winds that tore down trees, enforcing planes to be grounded for days on end. Instead, ground training and exercises were scheduled, along with lectures and talks by specialist officers and leaders. One of the lectures was to all Pilots and their engineers on the Mk 8 Auto-Control.

On the 13th January, Charlie was appointed to Commissioned Rank as Pilot Officer and discharged from General Duties.

Christmas Day presented Charlie with another promotion, as Temporary Warrant Officer, too late to include in his Christmas greetings already posted home. He'd sent an official card featuring the aircraft, BQ-B 'Phantom of the Ruhr' with RAF insignia and 'Greetings' dated 25/11/44. Whether they had flown in the Phantom or not, this aircraft, which would end up flying 121 operations was a hero to all the aircrews. Two other aircraft with 550 Squadron, BQ-F 'Press on Regardless' and BQ-V 'The Vulture Strikes' would also tally up more than 100 operations.

Once the weather conditions improved, the crews were back into the thick of it, with attacks focused on targets that played a vital part in the enemy's industry and transport: oil plants, chemical works, railway centres and transportation networks. Alongside other squadrons that were part of Bomber Command Group 1, Lancaster crews from 550 Squadron were deployed in ops to locations that would become familiar markers on the navigational maps and in WWII history.

Nuremberg was the target for a mission that Charlie particularly recalled:

> *We had trouble from start to finish. Even as we were taking off, lights came on, indicating the starboard wing tanks were empty, which turned out to be an electrical fault.*
>
> *Then when we got to Stuttgart, the German fighters picked us up and in the half hour it took to get from there to Nuremberg, they shot down 26 Lancasters.*
>
> *The fighters would line us up in their gun sights, knowing a Lancaster was 102 feet across. When the wings of the Lancaster filled the circle of their sights, they knew we were 800 to 1,000 yards away and that's when they'd shoot.*
>
> *I saw a lot of Lancasters shot down on that raid, and I know one was a Pathfinder because I saw the markers they were carrying explode.*
>
> *Then one of our engines began revving very high, shaking the plane like a leaf. You couldn't hear anyone speak, it was so loud. We got it a bit under control and even though we were having trouble from the flak, I suggested it was the lesser of two evils to fly lower and avoid the fighters.*

That night there was a risk of crossing paths with other aircraft, making it necessary for the pilot, George Cowper to take their Lancaster down to 8,000 feet from the usual range between

16,000 and 20,000 feet. Then one of the starboard engines had to be shut down after a hole was shot in the coolant. Eventually they flew back to England on three engines at 2,000 feet, arriving back to base quite late.

Many years later, Charlie met two Australians living in SE Queensland, both former air gunners. They remembered the raid. At that time they were prisoners-of-war in a camp outside Nuremberg, after their Lancaster had been shot down some time previously. They remembered on that particular night hearing a terrible screaming noise coming from one of the bombers.

CHAPTER 18

The Odds

50 Squadron had fought hard-won successes in bombing raids but not without suffering the loss of aircraft and good men, who were keenly mourned by all involved. Its losses were comparatively less than those suffered by other squadrons.

The Bomber Command statistics show huge losses. In total 364,514 operational sorties were flown and 8,325 aircraft lost in action. Bomber Command aircrews suffered a high casualty rate: of a total of 125,000 aircrew: 5,205 were killed (a 46 percent death rate), a further 8,403 were wounded in action and 9,838 became prisoners of war. Therefore, a total of 75,446 airmen (60 percent of operational airmen) were killed, wounded or taken prisoner. A memorial in Green Park in London was unveiled by Queen Elizabeth II on 28 June 2012 to highlight the heavy casualties suffered by the aircrews during the war.

Most aircrew were aged between 19 and 25, the average age being 21, although some were as young as 16, and one was in his sixties.

A total of 126 squadrons served with Bomber Command. Of these, 32 were officially non-British units: 15 RCAF Squadrons (Canada), eight RAAF squadrons (Australia), two RNZAF squadrons (New Zealand), four Polish squadrons, two French squadrons and one Czechoslovakian squadron.

Towards the final stages of the bombing raids, the RAF gave crews calico flags of the Union Jack with the lettering, "*I am an Englishman*", written on them in Russian. Charlie related that the "top brass" was concerned if crews were shot down, the Russians might be more likely than the Germans to kill the Allied airmen.

For those returning safely from raids, the overwhelming emotion was relief at having survived.

They were welcomed back by ground staff and a generous supply of tea, hot coffee and rum, usually dispensed by the Unit Padre. This helped to some extent in restoring a degree of comfort and sanity. Another compensation was the hearty breakfast of bacon and eggs, enjoyed by few people at that time of war rationing. A hot meal was followed by sleep in a relatively warm and comfortable bed in the wooden huts offering far better conditions than on the aircraft. They could sleep more easily when there was the possibility of no flights for a few days.

RAF stations generally had good leisure facilities and to relieve the stress of combat, there were frequent dances, mess parties and variety shows. Although scheduled leave was likely to be cancelled or shortened with changing circumstances, aircrew were given generous amounts of leave. Every six weeks they had a seven-day pass and shorter periods granted during prolonged bad weather or after difficult operations.

Finally, the Nuremberg sortie on 16 March 1945 in the aircraft PD320 "H" was the completion of tour for the George Cowper crew. Only a week later that same aircraft, PD320 "H" would be listed as Failed to Return (FTR) with nothing heard from the aircraft after take-off, the crew missing and then recorded as Killed in Action (KIA). This plane had taken the Cowper crew on at least 13 sorties and safely returned.

Flying Officer Charles Hume Baldwin

Charlie recalled the many sorties that had taken them over the whole length and depth of Germany, to Chemnitz, Pforzheim, Freiburg, Merseburg, Ludwigshafen and Aschaffenburg, to name a few, as well as the more familiar names associated with WWII history: Dresden, Cologne and Nuremberg. The names of cities and targets, no longer a dot or mark on a map, became a memory of danger, destruction, death – and the survival of those who made it back.

Sorties ranged in duration from five to more than ten hours. Danger was with them all the way: seeing their aircraft enmeshed by searchlights with little, if any, chance of escape and being obliterated by ground fire, or being shot down by fighters, as well as the odd mid-air collision along the way. On more than one occasion, luck would be with them, spared by a maladjustment of the enemy fighter's guns or a rookie pilot's inexperience, as indicated by the tracers.

On a sortie to an industrial target near Essen, George Cowper and his crew pressed on, with the aircraft's rear turret practically unserviceable throughout the trip. On another occasion, towards the end of tour, the crew was 200 miles short of the target when an engine caught fire and had to be feathered, but they completed the bombing operation and returned safely. On their last trip, the aircraft flew back on three engines, late back to base, but return they did. Charlie appreciated how lucky the George Cowper crew were in returning from their quota of missions:

> *We were ever so fortunate to be the "one" in the "one-in-three" equation of completing a tour of 30 ops. Our*

pilot, George Cowper was a nice gentlemanly type whose speech was always a pleasure to listen to. Much more importantly, he was an excellent pilot; highly skilled and always in control of any situation. He was a source of great confidence to the crew. I never knew him to raise his voice or see him affected by pressure.

The biggest risk period for not returning from an op, was during the first five and again after passing the 20 mark, when hope of surviving to finish the tour, could creep into their horizon. Determination, experience and a certain amount of luck played an important part in a crew doing its utmost to make it through. Unfortunately, many an airman was killed on his very first or last sortie.

On completion of their tour, the George Cowper crew was disbanded after being together for five intensive months, with them beating the odds. Most crews were lucky to get past three months.

In Charlie's memory, he had completed more than the quota of 30 operational sorties, 28 night and 5 day ops, but according to the navigator's logbook detailing 37 ops, it may have been more than 33. In the records 37 ops were listed, of which Charlie was absent for one of those, with another bomb-aimer replacing him on a raid to Kassel.

The way a tour was measured varied at times to include points per raid, but 30 ops was the usual number required in the latter stage of the war. Aircrews would often choose to stay together until all of them had completed their quota, with some doing extra ops in order not to break up the group.

After a few days at Killingholme Station Sick Quarters, Charlie was posted to 1656 Conversion unit before going to RAF Finningley Bomber Command Instructor School, a base in Yorkshire, where he was assigned to a Polish crew. He recalled this experience causing him some anxiety:

> I was more nervous flying with them over England than I was on operations. These blokes were a nightmare to go out with. I was supposed to be showing the bomb aimer what to do but they never stopped talking in Polish. It was impossible to teach the fellow much and it was nearly all in sign language. That was the last duty I had and I was glad when it was finished.

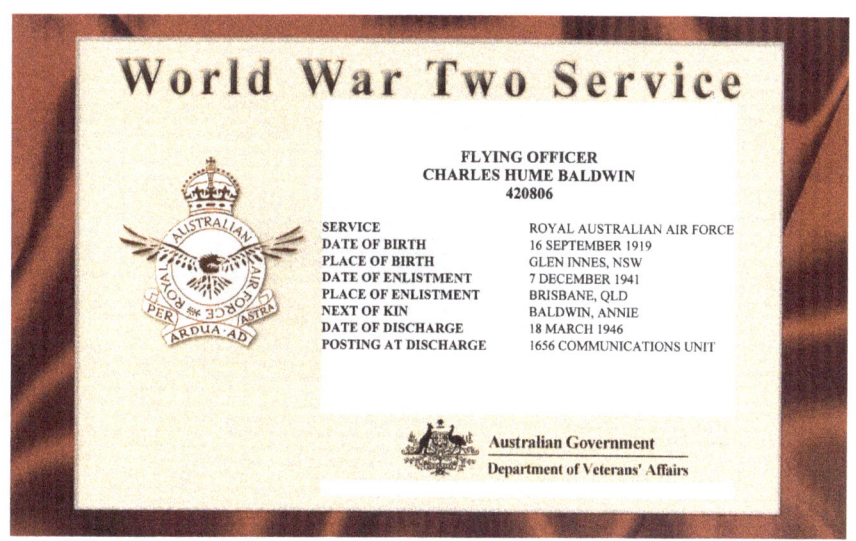

With the war declared over, Charlie could return home, his voyage from England embarking on 18 June 1945, Six weeks later he disembarked in Brisbane on 28 July, promoted during this time to Flying Officer. From his attachment to No.3 Personnel Depot, he was granted 106 days of leave, with a free rail pass around Queensland. On 18 March 1946, Charlie was demobilised. The duration of his war service had been four years and 102 days.

Charlie welcomed home from war by his sister, Nellie.

CHAPTER 19

Lost

After his return, Charlie's luggage was stolen, including his logbook, a loss he felt keenly, a companion taking him every step of the way. There was also a letter signed by Air Chief Marshall Sir Arthur Harris (Bomber Harris as he was known). As far as Charlie knew, the letter thanking him for his contribution to the war effort was not common. He'd only come across one other person with this letter of recognition.

Years later without his logbook, Charlie couldn't confirm some dates in recalling his experiences He wasn't certain of the date of a particular sortie, which he remembered in relation to an event more than 50 years after the war.

His memory was activated by the plane crash EgyptAir Flight 990 when it nose-dived into the Atlantic Ocean on 31 October 1999. Investigations were inconclusive as to the exact cause of the airliner plummeting into the sea. This incident reminded him of the plane he'd watched take a nose-dive into the sea during a mission to Germany. Thinking back, Charlie recalled:

We were one of a large force of bombers assigned to attack a target well into Germany. As we were crossing the coast of Kent on the way out, without a cloud in sight the sun was still shining with about half an hour or less to sunset and visibility was unlimited.

At this point it was my duty to attend to putting the bombsight in readiness for use at the target. In doing so the bombs were also rendered "live", ensuring that when they fell from the aircraft a small propeller, which protected the detonator in the nose of the bomb from operating until it was removed on its downward path by the airflow, they would be activated accordingly.

To perform the routine it was necessary to adopt a prone position. Immediately upon standing erect I looked out of a small window on the starboard side of the aircraft, when I saw another Lancaster bomber, probably about forty or fifty yards away standing directly nose down vertically with the engines running and heading into the sea. The aircraft disappeared completely and left a pattern on the surface as though the water was boiling. This occurred within a few seconds. Neither of the gunners saw it but that was not surprising. As we were travelling about 165mph there was no semblance of opportunity to see if anything came to the surface afterwards.

It appeared from the look of the propellers that the engines were functioning normally. From the close proximity and perfectly clear vision there was no suggestion of a collision. This instantaneous happening, so unlikely at that time and place, had a more chilling effect than seeing one shot down in action when one always knew and accepted that it would happen in enemy territory.

There was no question of dwelling on the matter. A constant lookout for the prevention of a collision was a requirement. The next concern was the next turning point which was almost universally at the French town of Abbeville at the mouth of the Somme, when a turn to port for the next leg of the journey commenced .

Over the following weeks and perhaps months I sought the views of senior pilots who had considerable operational experience, about the possibility of the automatic pilot being responsible for the tragedy, none of whom could or possibly would offer any suggestion as to the likelihood of the automatic pilot being responsible.

Our pilot, George Cowper was somewhat loathe to use the autopilot and it wasn't until well into our tour of operations that he "chanced" to use it and then only after we had dropped the bombs and had got away into an area of "open ground" where the presence of enemy attack had dropped off.

When the autopilot was engaged it was evident to everyone in the crew as the aircraft began to "wander" in a slow somewhat lazy and usually downward direction to begin with, but not with any undue sudden movement. From what I heard, I have the feeling pilots may have been reluctant to use it in case of fighter attack when precious seconds could have been lost between automatic operation and personal control of the aircraft.

It crossed Charlie's mind that the auto-pilot could have had some bearing on the similar fate of the Egyptian airliner crash, although different in some major aspects and in greatly differing circumstances. However, from information in the news, Charlie surmised there hadn't been any official inquiry as to the auto-pilot playing a part in the EgyptAir crash.

CHAPTER 20

Post War Life 1946

In June 1946, Charlie became engaged to Valma Grace Keating, only daughter of Len and Ethel Keating who were both well-known in the Beaudesert community. Later that year they married on 26 December, the wedding taking place at the Methodist Church, two doors from the family home where Valma and her three brothers shared a happy childhood. Charlie's family were still living at the Baldwin family property, Merton about 15 miles out of Beaudesert towards Rathdowney on the Mount Lindsay Highway.

After honeymooning in North Queensland, married life began in Innisfail NQ, where Charlie began work in the Queensland Agricultural Bank, established in 1902 and operating primarily as a rural financier. Through his work in the Innisfail district, doing property inspections as required to assess loan applications by farmers, Charlie's connection to the land was re-established.

One particular case lead to a friendship with an Italian man, Carlo Cottone who had left his Sicilian home in 1924 to escape

Wedding Day 26 December 1946 Beaudesert

the Mussolini regime. He settled near the small town of Babinda, where the rain was measured in feet, not inches. Carlo worked hard, cutting cane before buying his own land. Years later he was granted a loan through the Rural Assistance Scheme, whereby Charlie carried out an inspection of Carlo's property.

The two men shared an affinity with the land, but even so, Carlo wondered a few years later, why Charlie would leave his bank job to take on dairy farming. To Carlo's way of thinking,

Charlie holding baby Cheryl, Val in the government car provided by the Agricultural Bank, Innisfail, NQ

it didn't make sense to leave a secure job for the hard work of farming. But living on the land was in Charlie's heart and Val had fond memories of holidays on her uncle's farm at Westbrook near Toowoomba, so in 1952 they purchased a dairy farm at Millaa Millaa on the Atherton Tablelands. The reality of farming life being so demanding hadn't featured in the young Valma's holidays.

Hard work on the land brought Charlie contentment, at least to some extent. Days started early and ended with milking, riding bareback to round up the cows, accompanied by his faithful dog, Bounce. The cans of creamy milk were carted from the dairy on a wooden sled pulled by a draught horse, down to the road and sent to the butter factory. In between milking, it was a never-ending battle to improve pastures, slashing lantana, a shrub with pretty flowers and fast growth that threatened to take over large areas of land.

The arrival of daughters, Cheryl (1948) and Dianne (1954) gave Charlie and Val much joy, but happy childhood memories for the girls were tinged with the underlying current of Charlie's suffering. In quiet moments, a cloak of sadness came over him, perplexing to them at the time, not knowing what it stemmed from.

On Anzac Day, Charlie would attend the Dawn Service where his sad memories were brought to the fore. He remembered

Val with baby Dianne, Ian Arthy (son of Ena), Charlie, his sister Ena – at the farm Millaa Millaa 1954

those he'd served with, in particular his mates, Gus and Fred who had not survived the war and those he hadn't known, all of whom meant so much more than a name and service number. Reflecting on the loss of so many who hadn't returned home, he also thought about the many innocent people, who'd suffered in so many ways – the loss of life, homes, livelihood and those who bore permanent scars, both physical and mental.

His memories were numbed by a shot of rum in good strong coffee after the dawn ceremony, like the comfort experienced after a safe return from bombing missions.

The friendship between Carlo and Charlie continued and deepened with weekend family visits. Unexpectedly this was to have a significant impact on Charlie's post-war mental angst, which had been pushed deep in the dark recesses of his mind. From a conversation, sharing stories of their lives, something suddenly triggered painful memories of war and brought Charlie "unstuck".

With an overwhelming sense of responsibility for his part in the destruction of life, and as witness to the loss of fellow aircrew, feelings of guilt and grief erupted, unstoppable. Of deep importance to him was the maxim he aspired to live his life by – The Golden Rule: Do unto others, as you would have them do unto you.

Although not directly responsible, Charlie, as an air-bomber would feel remorse for the bombing of Italy, the home country of his good friend, Carlo. Undoubtedly he felt the injustice for Carlo's interment as an Italian POW, here in Australia,

Carlo's adopted country. Carlo had worked hard, becoming an Australian citizen and providing a new life for his wife, Elena and their three young sons. Without Carlo, Elena had been left to fight against opportunists endeavouring to take over management and ultimately ownership of their farm. Fortunately, Elena was a strong and wise woman, determined to remain living on the farm until Carlo returned to take up the reins.

Charlie's mental anguish resulted in being sent to Brisbane for "shock" treatment at the Greenslopes Repatriation Hospital with administration of a drug by injection, the purpose of which was not explained to him. Charlie remembered being questioned by the doctor treating him:

> *He asked me every single question as to my recollection of the experience of the aircraft crash in the Wellington bomber. It seemed to take a long time.*
>
> *It distressed me greatly as he required so much minute detail.*
>
> *If it was designed to be of benefit to me, it went horribly wrong because it had just the opposite effect. It revived all the worst memories of the whole affair and set me back considerably.*
>
> *At the end of it, the doctor was obviously aware of my very distressed condition when he said precisely, 'Well, you have just about had enough.'*

> *I was particularly incensed some time afterwards when the thought occurred to me that it might have been for the purpose of getting the truth from me. I couldn't imagine that would really be the case as surely all this would have been on record and in any case, why would I have any interest in distorting fact.*

Years later when Charlie was at retirement age, he documented his experiences in an application to the Department of Veteran Affairs for a pension. The DVA, founded in 1976 is responsible for delivering government programs for war veterans, the Australian Defence Force and Federal Police. In writing his account, Charlie realised it was the psychological effect resulting from the plane crash, compounded by witnessing fellow servicemen dying in similar circumstances, that had the greatest impact on his health. He made reference to his mental condition as Post Traumatic Stress Disorder, a term not widely known or understood at that time.

CHAPTER 21

Peace

After months of treatment in Brisbane, Charlie returned to the farm, which Val had been left to run with no help, not knowing when he was coming home. Some semblance of life went back to normal, busy days milking cows. Long heart-to-heart talks during visits from Carlo did more for Charlie's healing than the medical measures carried out in hospital.

Nightmares were still bound to happen, when he would wake up screaming. At other times, the horror of his suffering would surface unexpectedly. A fire occurring in a film would send Charlie racing from the picture theatre, panicked and upset. Seared in his memory was the smell of burnt flesh, blood, smoke and chemicals used in repairing aircraft. Throughout his life, he was obsessed with smoke detectors, fire alarms and escape routes.

As we grew older, my sister and I became aware of our father's service during the war, but knew none of the details that distressed him and affected his life on an on-going basis. In my memory, he had been a navigator during the war and wasn't aware of his role as bomb-aimer. My impression had possibly formed from his keen interest and training in navigation, which he would have mentioned rather than the bombing.

During this time Charlie found a new purpose, helping the immigrants on the farm next door. The Australian government's post-war plan to accept refugees and thousands of displaced people from camps in Europe was an agreement with the International Refugee Organization (IRO), formed in 1946. Immigration was one way to rebuild Australia's agricultural and industrial sectors. Under the Displaced Persons Program, Australia agreed to arrange transport, accommodation and employment for 4,000 displaced persons in 1947 and over a period of five years, accepted 170,000 displaced persons.

Every two years a couple or family from an Eastern European country would arrive to "share-farm" the dairy next door, while the owners lived a well-off, comfortable life in the city, many miles away. These "New Australians" came from Czechoslovakia, Hungary and Yugoslavia, working hard to make a new life, a safe life, far from conflict and persecution.

While Charlie helped them decipher the paperwork and check their rights with payment, Val would be asking for recipes, inspired by cheeses and sausages hanging in the kitchen. There were many happy meals shared and friendships formed over this period of time.

Not far away, another ex-serviceman was adjusting to post-war life, also on a dairy farm. There were frequent visits to see how David Beattie and his family were getting on. Charlie's life had shared some similarities with David's, although they came from different ends of Queensland. They had been born the same year (1919), one of seven children, both left school to work on the family farm during the Depression, their fathers had died the same year (1939), both from a heart attack and they had both become bomb-aimers serving with the RAF in England. Being attached to different squadrons, David with 467 Squadron (RAF Waddington, Lincs) and Charlie with 550 Squadron (RAF North Killingholme, Lincs), they hadn't known each other during the war.

David returned home to the Atherton Tablelands, fortunate to be alive but now blind. That didn't stop him taking on the challenge of dairy farming with the help of his wife, Mary. My childhood memory would be of their eldest child, a son about my age, guiding his father around the bails which had been painted white. Injured in the war, was all I'd been told.

Only recently I became aware of the circumstances leading to David's brush with death, when I began reading about the experiences of WWII airmen to gain a further understanding of that time. The graphic account of this event was a jolt when I recognised David's name.

He had been severely injured by a bullet through his head during an attack by a German night-fighter, damaging the aircraft. Despite his critical condition, David was still concerned about his task of releasing the bombs. Two of the crew administered what first aid they could, convinced he wouldn't live. They completed the bombing, with the aircraft still under attack, suffering further damage from continued attacks on the journey home. It was riddled with bullets, the radio not operating, only two engines still working and a different return course had to be navigated.

That the pilot and crew made it back at all was an indicator of the heroism Charlie had earlier referred to.

CHAPTER 22

Legacy

Another legacy of war would be the widows and children without husbands and fathers. Charlie joined the RSL (Returned and Services League, working with Legacy, a not-for-profit organisation established in 1923. Its aim was to care for the dependents of deceased Australian service men and women. As a Legatee, Charlie would provide support for a local family, ensuring they received the entitlements and government benefits and that the children would receive a good education with financial help from Legacy.

Fun events were organised annually, one being at Christmas with a huge tree, the arrival of Santa and receiving a gift, usually a book. Our family went with the Legacy family to beach picnics and the Christmas party. I recalled these happy occasions with Sandra, the daughter who happened to be in my class at school.

When time allowed, Charlie enjoyed listening to music and radio programs. War-time music evoked a sense of nostalgia, the camaraderie of mates who shared the adventures, the good times, dancing like there was no tomorrow, which would turn out to be true for so many of them and the other side of the coin, the fear and courage to carry out their duty and deal with the aftermath. The Glenn Miller Band was a favourite of Charlie's, as too the songs by Vera Lynn: *We'll Meet Again* and *As Time Goes By*.

As time went by, farm life became too much work, complicated by Charlie's inner torment affecting both his and Valma's health, so the farm was sold and Charlie returned to the Agricultural Bank with the family moving to south-west Queensland. By distance education, he studied a course to become an Associate of the Commonwealth Institute of Valuers working for the Qld Government Valuer General's Department.

This job meant driving long distances to remote properties in western Qld. He'd often camp for the night by the roadside, light a fire, boil the billy for a pannikin of strong black tea and cook damper with spuds thrown on the coals before sleeping in his swag under the stars. At peace with nature, Charlie enjoyed being close to the land.

Occasionally Charlie took a flight, courtesy of the property owner in a single-engine Cessna, the easiest way to cover distances and check out the terrain of properties that stretched over huge areas. With the pain of war a distant memory, a joy of flying was still with him.

Sadly Charlie and Val's marriage had been tested by the aftermath of war, which changed their happy, in-love relationship starting out together, ending shortly before their 25th silver anniversary. However, they remained on friendly terms, later residing in the same general area along the Fraser Coast, Queensland. Val had remarried some time previously and moved from Sydney with her husband, Alan Lyon after his retirement.

On Charlie's retirement he again settled on the land. He bought a small property between Gympie and Tin Can Bay, where a few cattle, a horse and a dog kept him company, not to mention a snake or two. Enjoying a more relaxed life, Charlie's sense of humour resurfaced, not being entirely extinguished by the toll of war. There was much hilarity in recounting the tussle Charlie had with one snake, which involved pots of boiling water, his favourite golf club and locking the "not happy" snake in the wardrobe until the next morning.

Correspondence between Charlie and his sisters was a constant throughout their lives, writing letters enclosed with funny snippets and cartoons cut out of the newspaper. Getting together on occasions, they would share a beer and hearty chuckles over some little story that tickled their fancy.

Charlie's post-war reflections revealed how much he had missed his family during the war and the changes in adapting to a social life, so different to his life at home, so far away. The distance was accentuated by the limited correspondence from his family. Looking back, he recalled:

> *At the time I went to Canada under the EATS training, I had neither tasted intoxicating liquor or smoked a cigarette. After a period of some months, I started smoking, principally or perhaps wholly because of the social structure and the inevitable confinement within places, such as dancing establishments, some picture theatres and buses, where it was impossible to escape the effects of smoking by others.*
>
> *There were times also that I felt disconsolate and dejected by lack of contact with my family, which had some effect upon my desire to smoke. Upon arriving in England, on experiencing and understanding the mood of the population at large, I easily fell into their way of life, in which smoking and perhaps to a lesser extent overall, drinking was accepted and practiced more universally, in my humble estimation, than either at home or in Canada.*

Smoking had become more than a social habit for Charlie, as it was for so many of the aircrews waiting to take-off on the bombing missions, but after the war, smoking distressed Charlie rather than settled his nerves. He quickly gave it up, but being in the presence of smokers would always cause him a great deal of discomfort. It didn't help that Val took up smoking, perhaps to settle her own nerves during tough days on the farm.

In reflecting on the war and the bombing over Europe, the "business end" of things, as Charlie once referred to it, he summed it up:

> *It was a life and death situation, with serious lifelong effects. However, peace brought everybody back to earth and a fresh start had to be made, and for many, changed circumstances had to be dealt with for the rest of their lives.*

POSTSCRIPT

A Letter from Cyril Fradley, Hixon UK, 1951

Without having any correspondence in Charlie's memorabilia that related to his time in the UK, to what extent Charlie had written home remains unknown. It would have been difficult, both physically and mentally for Charlie to write home after being injured in the plane crash at RAF Hixon. His main concern would have been to protect his family from knowing the full extent of his suffering. Later, when flying with 550 Squadron, much of Charlie's time would be focused on the intense bombing missions over Germany, with the dangers he faced not something he'd dwell on in correspondence home.

Many of the aircrew waiting for takeoff would pen a quick note home, giving instructions to a mate on what to do with it, if they didn't make it back from the mission. I wondered if Charlie wrote one of these last-minute letters to a girl in Toronto. During training in Canada, many servicemen met and fell in love, some of them returning to marry their sweetheart.

On the back:
Summer last year was during the last three days of June and here I'm enjoying it as much as possible.
Love from Charlie

In Charlie's official records, his mother Annie Baldwin was listed as next-of-kin. However, on two different forms that required a contact to be notified in case of causality, there were two names with the same surname, one being Mrs and the other Miss, with different addresses in Toronto. It was a detail that presented the possibility that during his training in Canada, Charlie had formed a close relationship with the younger woman, the older one being her mother. Not knowing anything about this prior to reading his official records, it was a curious revelation and remains a mystery.

At the time Charlie enlisted with the RAAF he was single and possibly hadn't met Valma, who had just turned 15 years old.

Having accepted that any evidence of Charlie's correspondence from England was lost, an unexpected discovery was made; three items – a photo, a card and a letter retrieved from boxes of letters and cards kept over the years by Charlie's sister Ena and later inherited by her elder son, Ian.

The photo shows Charlie relaxing in what appears to be a rocking chair. From what he had written on the back of the photo, he would have been referring to the previous year in Canada and likely to be on leave before departure for the UK, where he would later be enjoying the English sunshine. From the photo and his words, "as much as possible", it could be surmised that it was taken during the later stage of his convalescence following the aircraft crash at Hixon, Staffordshire. Charlie detailed in a later report that after the accident his mobility had been affected causing him difficulty walking, with damage also to his shoulder.

The postcard, dated 11th August 1943, was of Anne Hathaway's Cottage at Stratford-On-Avon, from an original Water Colour painting. Quintessentially English, the cottage featured thatched roofing, chimneys and the wattle and daub structure dating back to 1463, framed within the pretty cottage garden. On the back Charlie wrote that he was on seven days leave and had been to Shakespeare's place, he described as very nice. The next day he'd be going to London on the way back to camp.

Postcard from Charlie, 1943 – Anne Hathaway's Cottage, Stratford-On-Avon, Warwickshire

However, it was the letter that was of most interest. Dated 15 July 1951, the lengthy letter had been written by Cyril Fradley, the village baker in Hixon, Staffordshire replying to a letter he'd received from Charlie's mother, Annie.

A week earlier, Cyril had received an Airmail letter from Charlie, but he also expressed his pleasure in receiving news from Annie. Cyril wrote:

> *It gave myself and my wife the greatest satisfaction to know that Charlie had arrived home safe and sound and we as parents know only too well how you, his mother must have felt when he was restored to you.*

*CHARLIE – a journey from the Aussie
bush to battle skies far away*

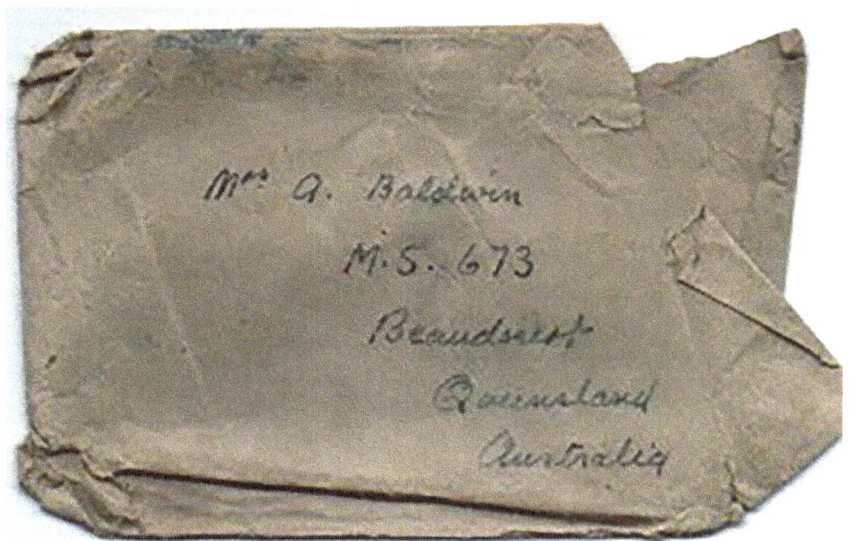

**Cyril Fradley's letter to Charlie's mother: Mrs A Baldwin,
M.S. (Mail Service), Beaudesert, Queensland dated 15 July 1951**

In her letter to Cyril, Annie indicated Charlie had said little about the day when he'd been injured escaping the aircraft crash. That fateful day in February 1944, when Cyril and Charlie were both injured by the flames of the burning aircraft, would scar them both for life and always connect them. Every day when he looked at his hands, Cyril would be reminded of his effort to save the life of Gus (Arthur Welstead) and Charlie would forever remember his own fortunate escape.

Cyril responded to Annie's unwritten question about that event:

> *Somehow I feel you would like to know, and again I feel you ought to know…*

In telling you how I saw it, don't think I am attempting to glorify anything I had to do with it. I only want you as a mother to try and see, what your son, as many another mother's son went through in the course of their duty.

It was a sunny Sunday morning as Cyril penned his letter to Annie, looking across to the dark shape of No. 5 Hangar spoiling his view, so readily reminding him of that tragic afternoon. In lengthy detail, he recalled what happened on that cold, blustery day, the wind almost gale force. Hearing the aircraft engine badly misfiring, he'd raced down the road, in time to witness the forced landing and crash, the aircraft

**Old Hangar on North Killingholme Airfield
Photo taken from Crook Mill Road.**

engulfed in fire in front of No. 5 Hangar, which was just nearing completion.

Cyril could see the crew scrambling out of dense smoke and flames to safety. Although he didn't know it was Charlie at the time, he later knew this after talking to him in the hospital. Cyril and others had watched Charlie's escape, which he described to Annie:

> *From the roadside we could see him very distinctly and then almost as if he was preparing to dive into a bathing pool, he threw himself into the surrounding ring of flames and smoke and that was how his face was so badly injured. Apparently he hit the already finished concrete approach to the Hanger very heavily but by a miracle he got out.*
>
> *As some of the others came to stand close by, with their clothes still smouldering, we could see Charlie's pal, Sgt Arthur Welstead [Gus] hanging by his parachute harness having been caught by the back on one of the flanged escape doors.*

Cyril did not go into any detail about his effort to rescue Gus, expressing to Charlie's mother, Annie that his biggest regret was that Gus did not live. Perhaps by the time Cyril wrote, he'd reflected on what that would have meant, had Gus lived. Perhaps with this in mind, he pointed out:

> *Believe me Mrs Baldwin, his burns were HORRIBLE.*

A few months after the rescue, Cyril's sorrow deepened, as he wrote it to Annie:

> *I went up to London to receive my award from the hands of H. M. the King. I met the family [of Gus Welstead]: the father, a rather thin, weary looking man who was a dock worker, the mother who was ill at the time, but who insisted on coming to say a big thank-you, and his sister, just on school leaving age.*
>
> *The father insisted I should tell them the whole story as I saw it. I did, slowly without any trimming. They heard me through in silence. How my heart bled for them, the agonised look on the fallen face of the father, the slowly welling tears in the mother's eyes, the tight shut lips of the sister, the strangled sob in her throat, I can hear them now.*
>
> *And so my tale is told. A lot of things that may have puzzled you since you have had Charlie back may now have their answer. In your letter you say the marks of his burns are gradually disappearing, that is a good thing and I do hope that with the passage of time the memory of those terrible days will be forgotten altogether.*

In his letter to Annie, Cyril indicated there was no need for Charlie to ever know that Cyril had told her about that day. He went on to say:

> *I am writing to him [Charlie] again shortly, but if it is that this letter reaches you first and you see him in the meantime give him a pat on the shoulder and a friendly hug.*

Cyril's letter took on a lighter tone, writing about other general matters. He mentioned the newspaper cuttings that Annie had sent as they were of interest to his 14 year old son, a keen cricketer, an interest that Charlie shared with him. At the same age, Charlie used to walk some distance from his home at Merton to a neighbouring farm to listen to the radio broadcast of the matches being played between England and Australia.

Then war had disrupted life and the cricket, the situation best summed up with a quote by Sir Home Gordon, who was an English Baronet, journalist and author, best known for his writing on cricket. In the September 1939 issue of the monthly cricket magazine, *The Cricketer*, he summed up the situation:

> *England has now begun the grim Test match against Germany.*

Cyril continued, telling Annie about his wife and their four children, their activities as well as explaining the English system of schooling. He also described the surrounding countryside, so Annie could picture where Charlie and Fred had cycled on those sunny afternoons, idling away time until they took to the skies again.

In 1977, Charlie took a ride down memory lane, returning to Canada, where he met up with Harold Graham, the crew's rear gunner. Sadly the crew's mid-upper gunner, Don Morrison had been killed in a motor bike accident. From Canada, Charlie again spent time in England where he would have seen George Cowper, the crew's pilot, who was living in London at that time. They had been in contact but the details of their meeting are unknown.

Charlie would notice vast differences from his wartime experience in England. The journey from London, northwest to the village of Hixon, more than thirty years down the track offered Charlie better transport than the bus, which used to take the aircrew from one RAF posting to the next, at least to places where there was no rail. Their luggage was sent separately on the top of a flat-top semi.

Arriving at Hixon, Charlie returned to the scene of the aircraft crash close to where Cyril the baker was still working and living. Undoubtedly there was a lot of reminiscing for Cyril and Charlie, the years falling away between that fateful day in February 1944 and Charlie's return, their memories revived over a cup of tea, or perhaps an English pint.

APPENDIX

Records - 550 Squadron database
Logbook of Stan Willis W/Op

Aircrew – Piloted by George Cowper

F/O G G H Cowper (P)
Sgt R H Whitehead (F/Eng)
F/O R M Mills (Nav)
F/Sgt C H Baldwin (A/B)
F/Sgt S E Willis (W/Op)
Sgt S E Morrison (MU/AG)
Sgt H R Graham (R/AG)

Between 31 October 1944 and 16 March 1945, the total ops flown by Cowper's crew: 29 Night & 7 Day

Ops flown by Charlie: 28 Night & 7 Day, although his recollection is 5 Day raids, but without his own Log Book it's difficult to be certain.

550 Squadron Aircraft – Call Sign: Baker-Queenie

Aircraft flown on Ops by George Cowper Crew

JB 345 BQ - C
ME 301 BQ - X
ME 390 BQ - A
NG 221 BQ - F
NG 290 BQ - L
NG 336 BQ - B
NG 390 BQ - H2 / J
PD 320 BQ - H

Charlie's Airman's Record Sheet
(Active Service – Overseas)

CHARLIE – a journey from the Aussie bush to battle skies far away

Bomber Command Medal

**Awarded to Charlie: 1939-45 Star
France and Germany Star
Defence Medal
War Medal 1939-45
Australia Service Medal 1939-45
Returned from Active Service Badge**

CHARLES HUME BALDWIN

16 September 1919 – 23 March 2005

Having celebrated his 85th birthday in 2004, Charlie passed away on 23 March 2005. His funeral was held at St Peter's Anglican Church, Gympie. A favourite hymn of Charlie's, *Jerusalem,* came from a poem written by William Blake and was set to music more than 100 years later, in 1916 by Sir Hubert Parry as a patriotic gesture to lift the spirits of people during the darkest days of World War I.

It would again have been a hymn sung during WWII and of comfort to Charlie, who developed a close affinity to England's green and pleasant land.

Charlie, rests in peace at Mt Thompson Memorial Gardens and Crematorium in the War Memorial Wall.

AUTHOR CHERYL BALDWIN

The author, Cheryl Ann Baldwin, a WWII post-war baby born in NQ, is the elder daughter of Charlie and Val (both deceased). Her childhood and primary schooling was spent on the Atherton Tablelands and secondary schooling at boarding school in Brisbane.

After years of living in far-ranging places, Cheryl is now Anna (Annaxue Yang) and lives in Toowoomba, Qld.

ACKNOWLEDGEMENTS

In writing my father's story of his journey to war and home, I'm grateful to family and friends for their interest and encouragement. In particular, thanks to my sister, Dianne Watts, our cousins John Eglington (son of Charlie's sister, Bonnie) and Ian Arthy (son of Charlie's sister, Ena) for being keepers of family history, especially the memorabilia and information relative to Charlie's war experience.

A large collection of information, including documents obtained from the National Archives, personal papers, photos and family history has been made available by John Eglington, which has made it possible to write Charlie's journey from Queensland to the war over Europe and home.

Many thanks to Ian Arthy for discussions on family history and discovery of the letter from Cyril Fradley, the village baker who witnessed Charlie's escape from the plane crash at RAF Hixon.

A huge thank you to my friends Hazel and Richard Kettle, Gloucestershire, for the photo inspiring me to write Charlie's story and the wealth of information researched by Richard available in the UK.

I express gratitude to Keira Quinn Lockyer, Ballarat U3a Writing Family History tutor for expert guidance in the initial proposal and draft of Charlie's story.

Thanks to my Ballarat friends, Debra Dickson and Clem Barnett for your interest and encouragement in zoom conversations, with added help from Clem in finding books for my research.

In searching for the family of Charlie's friend, Carlo, it's been a delight to reconnect with the Cottone family after losing contact many years ago. Unknown to me at the time, Davide had written and published the story of Carlo's life, including a conversation between those two friends, Carlo and Charlie. I'm particularly grateful to Davide for his friendship, enthusiastic encouragement and valuable feedback in my endeavour to tell Charlie's war story.

My friends, who have stepped into the role of Beta readers have also provided valuable feedback and support, which I greatly appreciate: fellow writers, Jen Gibson and Lyn McGettigan, both published authors of family memoirs; Narelle Moore, my life-long friend, for reading my writing, encouragement always and sharing childhood memories, growing up in those post-war years on dairy farms across the road from one another, close to the town of Millaa Millaa on the Atherton Tablelands.

My gratitude to Ian Whitaker, Millaa Millaa classmate with his own connection to the war experience of a family member, Bill Jackson who took a similar path joining the RAAF and served with Bomber Command. Also, Bill was a very close friend of

David Beattie, and with Charlie, they shared a bond forged from their war experience. With a depth of knowledge relating to all military history, Ian's reading and suggestions were valuable in the final draft of the story.

I'm indebted to my close friend, Ruth Drynan, our connections taking us from boarding school days and beyond as we continue to share our life journeys. Thank you Ruth, always for your encouragement and particularly for your keen eye and valued knowledge of the English language in proof-reading the manuscript before going to print.

RESOURCES

Charlie's war service records provide the dates, postings, promotions and places of the time he served in the RAAF, RCAF and RAF. However, it's from his own hand-written and sometimes typed recollections detailing the events which impacted his experience during the war, that the story of his journey emerges.

A timeline of missions flown by Charlie, as the bomb-aimer in the crew of George Cowper, pilot, have been accessed from a copy of the logbook of Stan Willis, wireless operator. The logbook includes a record of the mission dates, identification of the aircraft used by the crew, time of departure, proposed height for the flight and target to be attacked.

Other details of Charlie's story can be attributed to an interview in 1992 by Margaret Andrew, who wrote about his life, particularly his war service. Her article, much of it in Charlie's own words, appeared in two parts:

Remembering the brave young men from the RAAF; Leisuretime, Issue 06 - 19 June, 1992
Memories of World War II Leisuretime, Issue 20 June - 03 July 1992

Leisuretime was a community magazine distributed in Gympie and the Cooloola regions of SE Qld, where Charlie lived after his retirement in 1980.

A dedicated group of people maintain a comprehensive database of records online at the website of 550 Squadron Association, North Killingholme, Lincolnshire: www.550squadronassociation.org.uk

This information is made available to the families and friends of the squadron service personnel and continues to be updated, providing details of the bombing operations undertaken by the many men and women who served with 550 Squadron. The Association was set up in 1991, and since then annual commemorations and reunion events continue to acknowledge the contribution and sacrifice made by these men and women, always remembered.

Regular newsletters and correspondence keeps the Australian branch of the Association updated, along with those living in other countries. Charlie and Stan Willis, the Wireless-Op had both joined. Plans to attend a reunion in 2002 were cancelled for Charlie, due to his doctor's advice.

> *Five Fifty – The wartime story of a Lancaster squadron, Patrick Otter, Newark, Nottinghamshire, England 2017.*

This excellent account is available from 550 Squadron Association, North Killingholme, Lincs. and details the squadron's story from its formation in November 1943 to October 1945. It paints the reality of daily hardships, the demands of bombing operations, the successes and losses suffered with photos of aircrews and ground-crews.

Research on official websites provides a wealth of information relevant to Charlie's experience. Numerous sources of reference include the following:

Lincolnshire RAF and Airfields History:
http://www.raf-lincolnshire.info/bombercommand/bombercmd.htm

The International Bomber Command Centre, Lincoln:
https://internationalbcc.co.uk/en.wikipedia.org/wiki/RAF_Bomber_Command_aircrew_of_World_War_II

Photo credit (Page 40): Nigel Ish
https://i0.wp.comwww.defensemedianetwork.com/wp-content/uploads/2010/06/Lysander_bank.jpg?ssl=1)

Photo credit (Page 116) Old Hangar on North Killingholme Airfield Photo taken from Crook Mill Road.
(Source: geograph.org.uk Attribution: David Wright July 2009)

Researching information and stories of Australian men, who joined the RAAF and served with the RAF can be found in numerous accounts. These books listed are only a few of those which took me on a journey of discovery, providing me with deeper insight into Charlie's experience. In addition, one book was of particular relevance to his friend and fellow serviceman, David Beattie.

Lancaster Men The Aussie Heroes of Bomber Command, Peter Rees, Sydney, 2013.

Reading this book, I learned the details of the operational sortie that would result in the damaged aircraft 'L for Love' limping home with the injured bomb-aimer, Flt/Sgt David Beattie, 467 Squadron RAF Waddington. Awarded the DFM (Distinguished Flying Medal) and although without his eyesight, David enjoyed a long and productive life with good health, humour and spirit (1919-2008).

Within Charlie's collection of information, there is an account written by the navigator on board the aircraft, 'L for Love': *Mission to the Dortmund-Ems Canal, 4 November 1944*, written by F/Sgt Ed (F.J.) Ward. It is Ed's own personal recollection that provides details of this bombing mission.

Don Charlwood wrote about his own WWII experience in his books:

No Moon Tonight, Don Charlwood, 1956, 2nd ed. 2000
Journeys into Night, Don Charlwood, 3rd ed. 2013

His accounts make compelling reading and are of particular relevance, with Charlie sharing a similar pathway, although a different timeline to Don, training in Canada, being stationed at RAF Lichfield, UK and with 103 Squadron RAF Elsham Wolds, UK.

Chased by the Sun: the Australians in Bomber Command in World War II, Hank Nelson, Allen & Unwin, Crows Nest, NSW 2006 edn.

In this book, the author provides a voice for the men who joined the RAAF during WWII and served with Bomber Command and relates a range of their experiences with a comprehensive depth to all aspects of their journeys. Nelson follows their stories from beginning to end and all the challenges faced in-between.

Bombing Germany: the Final Phase, Tony Redding, Barnsley, South Yorks, 2105.

The focus of this account is the bombing of the German city, Pforzheim, 23 February 1945, which was one of the operations flown by 550 Squadron. Charlie as part of the George Cowper crew took part in this sortie. Twelve aircrew contributed to the telling of this story, but the author has also presented the story of Pforzheim and the lives of individual residents, the men and women, who remembered their own youth, and what life was like for them and their families under Nazi Germany and the Allied bombing.

Crew The story of the men who flew RAAF Lancaster J for Jig, Mike Coleman, Sydney 2018.

This story was inspired by a memorial and a tall Kauri pine tree, tucked away in a Brisbane suburb, where the navigator, Sgt Clifford Hopgood had lived with his wife and young son.

There are four headstones, one of them engraved with Cliff Hopgood's name in the French village, Villers-Sous-Preny, where in 1944, the burial of these four 'Jig' aircrew was attended by thousands of French people in defiance of the Germans.

This story represents many of the connections made from a fateful event and date, that survive the distance of time and locations.

Lest We Forget

www.ingramcontent.com/pod-product-compliance
Lightning Source LLC
Chambersburg PA
CBHW040307170426
43194CB00022B/2928